Principles of Heat and Mass Transfer

Principles of Heat and Mass Transfer

Annika Brayan

Larsen & Keller
www.larsen-keller.com

Principles of Heat and Mass Transfer
Annika Brayan
ISBN: 978-1-64172-131-8 (Hardback)

Larsen & Keller
Published by Larsen and Keller Education,
5 Penn Plaza,
19th Floor,
New York, NY 10001, USA

Cataloging-in-Publication Data

Principles of heat and mass transfer / Annika Brayan.
p. cm.
Includes bibliographical references and index.
ISBN 978-1-64172-131-8
1. Heat. 2. Mass transfer. 3. Heat--Transmission. I. Brayan, Annika
QC254.2 .P75 2019
536--dc23

Printed and bound in China.

For more information regarding Larsen and Keller Education and its products, please visit the publisher's website www.larsen-keller.com

Table of Contents

Preface

Heat transfer is a sub-field of thermal engineering, which deals with the generation, conversion, use and exchange of thermal energy between physical systems. The fundamental mechanisms of heat transfer are conduction, convection, advection and radiation. It is crucial for phase transition in a thermodynamic system from one state of matter to the other. Heat transfer has wide applications in insulation, thermal management of electronic devices and systems, materials processing, etc. Mass transfer refers to the net movement of mass from one location to another. It may occur due to the processes of precipitation, absorption, evaporation, distillation, etc. Mass transfer is used widely in separations engineering, reaction engineering, heat transfer engineering, etc. This book is a valuable compilation of topics, ranging from the basic to the most complex theories and principles in the field of heat and mass transfer. Different approaches, evaluations, methodologies and studies have been included in this book. It aims to serve as a resource guide for students and experts alike and contribute to the growth of the discipline.

A detailed account of the significant topics covered in this book is provided below:

Chapter 1- Heat transfer is a discipline of thermal engineering. It is concerned with the conversion, generation and exchange of thermal energy between physical systems. This is an introductory chapter, which will introduce briefly all the significant principles of heat transfer and heat transfer physics and includes significant topics like heat-transfer fluid and thermal radiation.

Chapter 2- Heat is transferred within a body due to microscopic collisions between particles and electrons. This is termed as heat conduction. All the diverse aspects of heat conduction, such as heat equation, Fourier's law for heat conduction, thermal diffusivity, thermal conductivity, thermal effusivity, etc. have been carefully analyzed in this chapter.

Chapter 3- Convection is the heat transfer in fluids that occurs due to bulk transfer or movement of molecules within the fluid. Convection can be natural, forced or mixed where the two occur in combination. This chapter closely examines the key concepts of convection and its different mechanisms to provide an extensive understanding of heat transfer by convection.

Chapter 4- Heat can be transferred between two or more fluids using a device called a heat exchanger. It can be of different types such as regenerative heat exchanger, plate and frame heat exchanger, shell and tube heat exchanger and dynamic scraped surface heat exchanger, among others. This chapter has been carefully written to provide an easy understanding of heat exchanger and its varied types.

Chapter 5- Mass transfer is the movement of mass from one location to another through the processes of evaporation, absorption, distillation, precipitation, etc. The topics fundamental to the study of mass transfer, such as mass transfer coefficient, mass conservation, mass balance and mass diffusivity, have been discussed in great detail in this chapter.

It gives me an immense pleasure to thank our entire team for their efforts. Finally in the end, I would like to thank my family and colleagues who have been a great source of inspiration and support.

Annika Brayan

1

Introduction to Heat and Heat Transfer

Heat transfer is a discipline of thermal engineering. It is concerned with the conversion, generation and exchange of thermal energy between physical systems. This is an introductory chapter, which will introduce briefly all the significant principles of heat transfer and heat transfer physics and includes significant topics like heat-transfer fluid and thermal radiation.

Heat

Heat is referred to energy that is transferred from one body to another as a result of difference in temperature.

The picture above shows that Deserts are very hot during the day. Plants and animals living in the desert need special adaptations to survive in the harsh environment

An object decreases its temperature by releasing energy in the form of heat to its surroundings. And an object increases its temperature by gaining energy in the form of heat from its surroundings. The higher the temperature of an object is, the greater the tendency of that object to transfer heat. A substance may absorb heat without an increase in temperature by changing from one physical state to another, from solid to liquid (melting) from solid to vapour (suppression), from liquid to vapour (boiling), or from one solid form to another (crystalline transition). As stated heat is the transfer of energy and the transfer of energy always takes place from the region of the higher temperature to the region of the lower temperature. Heat is the energy in flow due to the difference in temperature.

Example: Consider a hot mug of tea. Let's assume that the initial temperature of the tea is 80°C. If we do not drink the tea and resist it, it will eventually reach room temperature and the tea will become cold over time. Here, the heat gets transferred from the tea and the mug to the surroundings which are at lower temperature than the tea until equilibrium is reached between tea and its surroundings. We refer to this transfer of energy from the tea and the mug to the air as Heat. Heat is the transfer of energy from a hot object to a colder object.

Heat Measurement Units

There are Three Main Units of Heat:

- Calorie – It is defined as the amount of heat needed to be transferred to increase the temperature of 1 gram of water by 1°C. It is represented by "Cal."
- British thermal unit (BTU) – It is defined as the amount of heat needed to be transferred to increase the temperature of one pound of water through 1°F. It is represented by "BTU."
- Joules – Joules is the SI unit of heat. It is equal to the work done when a force of one Newton is applied for a distance of one meter. It is represented by "J."

Heat Capacity

Heat capacity gives a measure of the quantity of heat absorbed by a system. It is defined as the amount of heat required to rise the temperature of the system through 1°C. The quantity of heat required to raise the temperature of a unit mass of any substance through 1°C is termed as the specific heat of the substance. Whereas, when the heat capacity is taken into account the whole mass of the substance is considered.

Heat Capacity Formula

Two types of heat capacity are used depending on the path followed by the reaction. If the change occurs at constant pressure it is expressed as Cp and if the change occurs at constant volume, it is expressed as Cv. Mathematically,

$$Cp = \frac{dqp}{dT} = \frac{dH}{dT}$$

$$Cv = \frac{dqp}{dT} = \frac{dE}{dT}.$$

The difference between Cp and Cv can be found from n CpΔT when pressure Cp is constant and n CvΔT when Cv volume is constant.

Since H = E+PV

$$\frac{dH}{dT} = \frac{dE}{dT} + \frac{d(PV)}{dT}$$

Heat Capacity Formula

$$Cp = Cv + \frac{d(PV)}{dT}$$

For solids and liquids, $\frac{d(PV)}{dT}$ is small and so Cp = Cv. For ideal gases Cp = Cv + R.

Calculating Heat Capacity

For an ideal gas the average kinetic energy would be equivalent to 3/2 R T, where KE avg represents the average random translational energy for 1 mole of gas at the given temperature T (kelvins). The heat capacity ratio or heat capacity can be calculated with more accuracy under ideal conditions.

$$E = \frac{3}{2}RT.$$

But dE/dT is the increase in energy per degree per mole or heat capacity at constant volume. Thus, Cv = dE/dT or 3/2 R for a mono atomic gas.

The calculation of heat capacities of ordinary mono atomic gases is given by the classical equipartition of energy principle. When the temperature is raised at constant pressure of 1 atm, the gas expands from V_1 to V_2. An additional energy is required to account for the expansion of the gas which is apart from the energy used to raise the temperature. As a result Cp must be greater than Cv. For an ideal gas this energy is easy to calculate.

The work done by expansion is the difference in initial and final PV product as PV has units of energy.

Thus, $P\Delta V = P(V_2 - V_1) = (PV_2 - PV_1)$

Since PV = RT for 1 mole of gas. The above equation becomes,

$P\Delta V = R\Delta T$ or $R = P\Delta V/\Delta T$

$$\text{Therefore, } Cp = Cv + R = \frac{3}{2}R + R = \frac{5}{2}R$$

$$Thus\ \frac{C_p}{C_v} = \frac{5}{3} = 1.67$$

The heat capacity ratio for di atomic gases is consistently less than 1.67. This is explained below. The heat capacity due to translational motion of the molecule Cv is equal to 3/2R. There are 3 independent velocity components associated with the translational motion, in three axes. If one motion is taken as contributing 3/2 R towards heat capacity, the other two contribute ½ R

Therefore, $Cv = \frac{3}{2}R + \frac{1}{2}R + \frac{1}{2}R = \frac{5}{2}R$

$$Cp = Cv + R = \frac{7}{2}R$$

$$\frac{C_p}{C_v} = \frac{7}{5} = 1.40$$

Hence the value of di atomic gases have lesser value of the ratio.

For mono atomic gases g = 1.67

For diatomic gases g = 1.40

For triatomic gases g = 1.30

Specific Heat Capacity Table

Specific Heat Capacity Ratios Cp/Cv for Some Mono Atomic and Di Atomic Gases:

Values in Cal mol-1K-1

Gas(Mono atomic)	Ratio	Gas(Di atomic)	Ratio
He	1.66	H_2	1.41
Ne	1.64	O_2	1.40
Ar	1.67	N_2	1.40
Kr	1.68	CO	1.40
Xe	1.66	NO	1.40
Hg	1.67	Cl_2	1.36

Heat Capacities of Some Solid Elements:

Cal $deg^{-1}mol^{-1}$

Element	At.Mass	Specific Heat	Heat capacity
Lithium	6.9	0.92	6.3
Beryllium	9.0	0.39	3.5
Magnesium	24.3	0.25	6.1
Carbon	12.0	0.12	1.4

Aluminium	27.0	0.21	5.7
Iron	55.8	0.11	6.1
Silver	107.9	0.056	6.0
Lead	207.2	0.031	6.4
Mercury	200.6	0.033	6.6

Heat Capacity of Water

Heat Capacity of Water and Specific Heat Capacity of Water:

Water has got a high specific heat capacity. It has the second highest value amongst all substances, the first being ammonia. Heat capacity of water is defined as the amount of heat required to raise the temperature of I g of water by 1 degree.

Note:Water absorbs an amount of heat before its temperature raises by a degree. It takes 4.184 J g^{-1} for water to raise the temperature by 10C. In comparison, Copper takes 0.385 J g^{-1} $degree^{-1}$. This property is attributed to the intermolecular hydrogen bonding that exists in water.

This can be experienced by the temperature of sea and the sand when both are exposed to the same hot Sun. Specific heat capacity of water is the heat capacity of water in its three states viz. Ice, Water and water vapour. The specific heat capacity of ice (solid) is 2.108 J g^{-1} $degree^{-1}$ and the specific heat capacity of water vapour (gas) is 1.996 J g^{-1} $degree^{-1}$. However, the specific heat content of water in its liquid form is 4.184 J g-1 or specific heat of water is 1 calorie/gram °C = 4.186 joule/gram °C.

Molar Heat Capacity of Water

Heat capacity of water is for 1 g of water. 18 g of water is 1 mole. Thus, the molar heat capacity of water is 18 X heat capacity of water.

Molar Heat Capacity

Molar Heat Capacities at Constant Pressure of Some Substances (C°p in J $mol^{-1}deg^{-1}$)

Gas	Value	Liquid	Value	Solid	Value
He, Ne, Ar, H	20.786	H_2O	75.291	C graphite	8.527
O	21.912	CH_3OH	81.6	Diamond	6.113
H_2	28.824	C_6H_6	136.03	S rhombic	22.72
N_2	29.125			S monoclinic	23.23
O_2	29.355			CaO	42.80

HCl	29.12			$Ca(OH)_2$	87.49
H_2O	33.577			Cu	24.435
CO	29.142			CuO	42.3
CO_2	37.11			Ag	25.35
C_2H_6	52.47			AgCl	50.79
C_6H_6	82.34				

Heat Capacity Problems

Example 1: A gas expands from a volume of 3.0 dm^3 to 5.0 dm^3 against constant external pressure of 3.0 atm. The work of expansion is used to heat 10.0 mol of water at temperature 290.0 K. Calculate final temperature of water. (Specific heat of water = 4.184 $JK^{-1}\ g^{-1}$)

Solution $W = - P (V_2 - V_1)$

$= (-3) \times (5 - 3) = (-6)$ L.atm.= $(-6) \times 101.25\ J = 6.07 \times 10^2\ J$

Now, Q = m. C.T

Final temperature

$= T + \Delta T = 290 + 0.8 = 290.8$ K

Specific Heat

When energy is given to a substance and the substance is not performing any work, it results into increase in temperature of the substance. (If the temperature of the substance is not increased, when the heat is given to it, and also the substance is not performing any work, it could lead to change in the phase of the substance and this phenomenon is termed as Phase Transition). Amount of heat, required to raise the temperature of a substance by a certain amount, depends on the properties of the substance and this amount of heat varies from substance to substance.

For example, the amount of energy or heat required to increase the temperature of 1 kg of water by 1°C is 4.186 J, but the amount of heat required to increase the temperature of 1 kg of copper by 1°C is only 387 J. There are two methods to increase the temperature of the substance:

1. By transferring heat or energy to it.
2. By doing Work on it.

Specific heat c of a substance is defined as the heat capacity of the substance per unit mass of the substance. Therefore, if energy Q is given to the substance having mass m, and it results in the change in the temperature of the substance by ΔT. Hence, Specific Heat of the substance is

$$c = \frac{Q}{m\Delta T}$$

Specific Heat Capacity

Heat capacity of a substance is denoted by C. It is defined as the amount of energy which is required to increase the temperature of the substance by 1°C.

According to the definition, if heat Q given to the substance increases the temperature of the substance by ΔT, then

$$Q = C\Delta T$$

Specific Heat Formula

Specific heat is essentially a measure of how thermally insensitive a substance is to the addition of energy. If the specific heat capacity of a substance is more, then for a given mass m of a substance, for a particular ΔT temperature change, more energy Q needs to be transferred to the substance as compared to the second substance having less specific heat capacity. Therefore, more the specific heat capacity, more energy needs to be transferred to the substance if the other conditions (Q, m,ΔT) are same.

Formula for specific heat can be written as,

$$c = \frac{Q}{m\Delta T}$$

Specific Heat Units : SI unit of Specific Heat is Joules per Kilogram Kelvin. (J/kg.K).

Specific Heat Ratio

The heat capacity ratio is the ratio of the heat capacity at constant pressure (CP) to heat capacity at constant volume (CV).

Specific heat Ratio = $\gamma = \frac{C_p}{C_v}$

Calculating Specific Heat

For calculating the specific heat of the substances we need to understand

Conservation of Energy: Calorimetry

One method to measuring the specific heat of the substance involves the steps of:

1. Heating a substance to temperature Tx.

2. Placing the substance in a beaker which contains water. Mass and temperature of the water are known.

3. Measuring the temperature of the water after thermal equilibrium is attained.

Small quantity of work is done in this process. According to the Law of the conservation of energy, the quantity of energy which leaves the sample of the substance (whose specific heat is unknown) is equal to the quantity of energy or heat which enters the water. This process is termed as the Calorimetry, and devices are called calorimeters.

According to the Conservation of energy Law, we can write

$$Q_{cold} = - Q_{hot}$$

which shows that total energy of the system remains constant, i.e., heat left by the substance is equal to the heat transferred to the water. The heat negative Q_{hot} shows that the hot substance is leaving the heat. The negative also signifies that the right side of the equation is positive and consistent with the left side of the equation. Left side is positive because heat is transferring to the water.

Take m_x as the mass of a substance sample whose specific heat we want to determine. Lets denote its specific heat y, c_x and Initial temperature by T_x. Similarly, let m_w, T_w and c_w, represents corresponding values for the cold water.

After mixing the substance in the cold water, energy transfer takes place and equilibrium is reached. Lets say, T_f is the Thermal equilibrium temperature after everything is mixed, then from the equation of specific heat capacity Q = mcΔT, we can equate the values on Left hand side and Right hand side. Substituting these values into above mentioned specific heat equation we get,

$$m_w c_w (T_f - T_w) = -m_x c_x (T_f - T_x)$$

Solving for C_X (Specific heat of the unknown substance we get)

$$C_X = \frac{m_w c_w (T_f - T_w)}{m_x (T_x - T_f)}$$

Latent Heat

The amount of absorbed or released heat is called latent heat. This heat is released or absorbed during physical change. It is named according to the process of physical change. It is expressed as per mole amount of released or absorbed heat taken by substance. It can be done with change in temperature or without change in temperature or phase change process. Let's discuss the detail description of latent heat and its different types during different stages.

But there are scenarios during which energy transfer between system and its surroundings does not result in the change in temperature of the system. This occurs whenever system changes from one of its phase to the another or we can say that when the transfer of energy results in the change in the physical properties of the system. This change is termed as Phase Change or Phase Transition. Two main Phase transitions are:

1. Solid to Liquid
2. Liquid to Gas

Other phase transition is when change takes place in the crystalline structure of the solid. When phase Transition occurs temperature of the system does not change but change in the internal energy of the system takes place.

At the boiling point of the water when heat is transferred to the water it results in the breaking of molecular bonds in water. As a result, the molecules become apart as their potential and kinetic energies get increased and water gets converted to the gaseous state. Every substance has different molecular and atomic structure. So, for the phase transition to occur every substance requires different amount of energy depending on their internal structure. It also depends on the mass of substance involved for the phase transition. The substance of a particular type with larger quantity and mass requires more heat for phase transition as compared to the other. It takes less energy to boil water in cooking pan than to boil a lake.

If a Q amount of energy needs to be transferred for phase change of a substance having mass m, the ratio symbolizes an important property of that substance. As the energy added or removed is not resulting in the change of temperature, the quantity L is termed as the latent heat. It can also be termed as the hidden heat of the substance.

Latent Heat for a Substance Depends Mainly on Two Factors :

1. Phase change Nature
2. Properties (atomic, molecular structure of the substance)

Latent Heat Formula

From the definition, it is clear that the amount of heat required for a phase transition is latent heat. The mathematical representation of the given statement is,

$$Q = mL$$

$$\mathrm{L} = \frac{Q}{m}$$

Where, L = Latent heat of the substance Q = Amount of energy transferred for the Phase change m = Mass of the substance

Latent Heat of Vaporization

Latent heat of Vaporization is the Latent heat (explained above) when phase of the substance changes from Liquid to gaseous state. Latent heat of vaporization is denoted by L_V.

Latent Heat of Condensation

Latent Heat of Condensation is defined as the quantity of heat needs to be transferred to convert a unit mass of a substance in gaseous phase into liquid phase at its condensation point at one atmosphere pressure.

$$\text{L} = \frac{Q}{m}\,\text{K} - \text{Cal/ Kg}$$

Latent Heat of Freezing

Latent Heat of Freezing is defined as the quantity of heat needs to be transferred to convert a unit mass of a substance in liquid phase into solid phase at its freezing point, at one atmosphere pressure.

$$\text{L} = \frac{Q}{m}\,\text{K} - \text{Calorie/ Kg}$$

Latent Heat of Fusion

Latent Heat of Fusion is the latent heat (explained above) when phase of the substance changes from Solid to Liquid phase. Latent heat of Fusion is denoted by L_F. Also fuse means melting.

For a particular substance Latent heat of Vaporization is generally greater than its Latent heat of Fusion.

Latent Heat of Melting

Latent Heat of Melting is defined as the quantity of heat needs to be transferred to convert a unit mass of a substance in solid phase into liquid phase at its melting point, at one Atmosphere pressure. It can be expressed as,

$$\text{L} = \frac{Q}{m}\,\text{Calorie/gm} \;\text{ or K} - \text{Calorie/Kg}$$

Latent Heat of Sublimation

Latent Heat of Sublimation is defined as the quantity of heat needs to be transferred to convert a unit mass of a substance in solid phase into gas phase without passing the substance through the intermediate state of Liquid, and directly to the gaseous phase at its sublimation point and at one atmosphere pressure.

Latent Heat Flux

Latent Heat Flux is denoted by Q_e when water evaporates, it removes significant amounts of heat. The energy that is added to break the hydrogen bonds to allow evaporation is removed by the water vapor. It can be calculated with the help of Bulk Formula,

$$Q_e = p_a L_E C_L U_{10}(q_a - q_s)$$

Where, p_a = Density Of air (~ 1.2 kg/m3)

L_E = Latent Heat of vaporization (~ 2260 kJ/kg)

C_L = Bulk transfer coefficient (Stanton Number)

U_{10} = Wind speed at 10 m

q_a = specific Humidity at saturation

q_s = specific Humidity of overlying air

Latent Heat vs Sensible Heat

Sensible Heat

When energy or heat is transferred to the substance, it results in increase in its temperature. The increase in the heat of the substance is called sensible heat. Similarly, when heat is transferred from the substance to its surroundings, heat is removed from the substance and results in decrease in the temperature of the system. The heat removed from the substance is also sensible heat. Heat, because of which change in temperature of the substance takes place, is called sensible heat.

Latent Heat

In contrast to Sensible Heat, Latent heat does not change the temperature of the system but it results in transition of the phase of the substance from one phase into other.

Latent Heats of Substances

Latent Heat of Water, Latent Heat of Ice and Latent Heat of Steam are given in the tables below. Latent Heat of Fusion and Latent Heat of Vaporization for different substances are given in the below tables.

Table:

Substance	Melting Point (°C)	Latent Heat of Fusion (J/kg)	Boiling Point (°C)	Latent heat of Vaporization (J\kg)
Helium	-269.65	5.23×10^{3}	-268.93	2.09×10^{4}
Nitrogen	-209.97	2.55×10^{4}	-195.81	2.01×10^{5}
Oxygen	-218.79	1.38×10^{4}	-182.97	2.13×10^{5}
Ethyl Alcohol	-114	1.04×10^{5}	78	8.54×10^{5}
Water	0.00	3.33×10^{5}	100.00	2.26×10^{6}
Sulphur	119	3.81×10^{4}	444.60	3.26×10^{5}
Lead	327.3	2.45×10^{4}	1750	8.70×10^{5}
Aluminum	660	3.97×10^{5}	2450	1.14×10^{7}
Silver	960.80	8.82×10^{4}	2193	2.33×10^{6}
Gold	1063.00	6.44×10^{4}	2660	1.58×10^{6}
Copper	1083	1.34×10^{5}	1187	5.06×10^{6}

Table:

Substance	Latent Heat of Fusion (kJ/kg)	Melting Point (°C)	Latent heat of Vaporization (J\kg)	Boiling Point (°C)
Alcohol,Ethyl	108	-114	855	78.3
Ammonia	339	-75	1369	- 33.34
Carbondioxide	184	-78	574	- 57
Helium			21	- 268.93
Hydrogen(2)	58	-259	455	- 253
Lead	24.5	327.5	871	1750
Nitrogen	25.7	-210	200	- 196
Oxygen	13.9	-219	213	- 183
R134a		-101	215.9	- 26.6
Toluene		-93	351	110.6
Turpentine			293	
Water	334	0	2260	100

Heat Transfer

Heat transfer is the process of transfer of heat from high temperature reservoir to low temperature reservoir. In terms of the thermodynamic system, heat transfer is the movement of heat across the boundary of the system due to temperature difference between the system and the surroundings. The heat transfer can also take place within the system due to temperature difference at various points inside the system. The difference in temperature is considered to be 'potential' that causes the flow of heat and the heat itself is called as flux.

Modes of Heat Transfer

There are three modes of heat transfer between the two bodies: conduction, convection and radiation. These have been described below:

Conduction: The transfer of heat between two solid bodies is called as conduction. It depends on the difference in temperature of the hot and cold body. Example of conduction heat transfer is two bodies at different temperature kept in contact with each other. Another example is heating one end of the metal like copper; due to conduction heat transfer the other end of the metal also gets heated.

Convection: The transfer of heat between the solid surface and the liquid is called as convection heat transfer. Let us considering a vessel of water being heated, in this case heating of water due to transfer of heat from the vessel is convection heat transfer.

Radiation: When two bodies are at different temperatures and separated by distance, the heat transfer between them is called as radiation heat transfer. In case of the conduction and convection heat transfer there is a media to transfer the heat, but in case of the radiation heat transfer there is no media. The radiation heat transfer occurs due to the electromagnetic waves that exist in the atmosphere. One of the most important examples of radiation heat transfer is the heat of the sun coming on the earth.

Heat Transfer as Per Second Law of Thermodynamics

As per the second law of thermodynamics the transfer of heat takes place from the body of high temperature to the body of low temperature. There won't be spontaneous transfer of heat from the body at low temperature to the body at high temperature. For heat transfer from low temperature body to high temperature body, external work has to be done.

The heat gained by the system or body is considered to be positive and the heat lost by the system is considered to be negative for the mathematical calculations. This implies that the heat flowing into the system is positive and heat flowing out of the system is negative. The amount of heat transfer is denoted by symbol Q.

The process in which there is no transfer of heat between the system and the surrounding is called as adiabatic process. The wall or boundary which does not allows the flow of the heat between the system and the surrounding is called as adiabatic wall and the wall that allows the flow of heat between the system and the surrounding is called diathermic wall.

In MKS system the unit of heat transfer is cal and in SI system it is Joule. The rate of transfer of heat is given in KW.

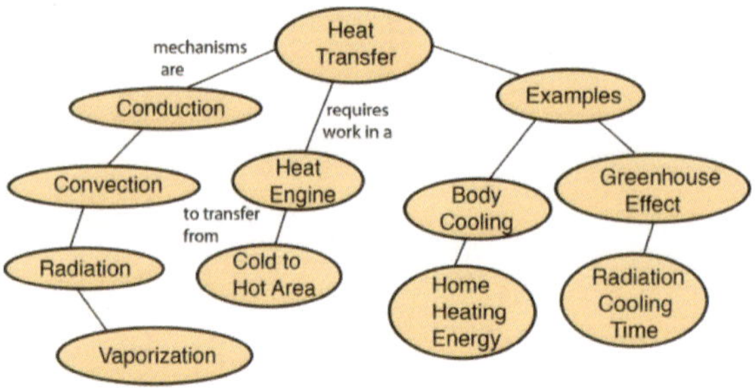

Heat Transfer Coefficient

Heat transfer coefficient is a quantitative characteristic of convective heat transfer between a fluid medium (a fluid) and the surface (wall) flowed over by the fluid. This characteristic appears as a proportionality factor a in the *Newton-Richmann relation*

$$q_w = \alpha(T_W - T_f),$$

where $q_w = -\lambda(\partial T/\partial n)_W$ is the heat flux density on the wall, T_w the wall temperature, T_t the characteristic fluid temperature, e.g., the temperature T_e far from the wall in an external flow, the bulk flow temperature T_b in tubes, etc. The unit of measurement in the international system of units (SI) is W/(m²K), 1 W/(m²K) = 0.86 kcal/(m²h°C) = 0.1761 Btu/(hft²°F) or 1 kcal/(m²h°C) = 1.1630 W/(m²K), 1 Btu/(hft²°F) = 5.6785 W/(m²K). The heat transfer coefficient has gained currency in calculations of convective heat transfer and in solving problems of external heat exchange between a heat conducting solid medium and its surroundings. Heat transfer coefficient depends on both the thermal properties of a medium, the hydrodynamic characteristics of its flow, and the hydrodynamic and thermal boundary conditions. Using the methods of similarity theory, the dependence of heat transfer coefficient on many factors can be represented in many cases of practical importance as compact relations between dimensionless parameters, known as similarity criteria. These relations are said to be generalized or similarity equations (formulas). The Nusselt number Nu = αl/λf or the Stanton number $St = \alpha/\left(\dot{m} C_{pf}\right)$ is used as a dimensionless number for heat transfer in these equations, where 1 is the characteristic dimension of the surface in the flow, $\dot{m} = \rho u$ the mass velocity of the fluid flow, λ_f and C_{pf} the fluid thermal conductivity and heat capacity. When solving the problems of heat conduction in a solid, the distribution of heat transfer coefficient α between the body and its surroundings is often given as a boundary condition. Here, it is useful to use a dimensionless independent parameter, the Biot number Bi = $\alpha l/\lambda_s$, where λ_s is the thermal conductivity of a solid and 1 its characteristic dimension. The dependence of the Nu and St numbers on the Re and

Pr numbers plays an essential role in heat transfer by forced convection. In the case of fully developed heat transfer in a circular tube with laminar fluid flow the Nusselt number is a constant, namely Nu = 3.66 at a constant wall temperature and 4.36 at a constant heat flux. In the case of free convection, the Nu number depends on the Gr and Pr numbers. When the heat capacity of the fluid varies substantially, the heat transfer coefficient is frequently determined in terms of enthalpy difference ($h_w - h_f$). The concept of heat transfer coefficient is also used in heat transfer with phase transformations in liquid (boiling, condensation). In this case the liquid temperature is characterized by the saturation temperature T_s. The order of magnitude of heat transfer coefficient for different cases of heat transfer is presented in table below.

When analyzing internal heat transfer in porous bodies, i.e., convective heat transfer between a rigid matrix and a fluid permeating through it, use is often made of the volumetric heat transfer coefficient

$$\alpha_V = q_v / (T_W - T_f),$$

where qv is the heat flux passing from the rigid matrix to the fluid in a unit volume of a porous body, T_w the local temperature of the matrix, and T_f the local bulk temperature of the fluid.

It should be emphasized that the constancy of α over a wide range of $\dot{q}_W$ and ΔT (other conditions being equal) is encountered only in the case of convective heat transfer when the physical properties of fluid change only slightly during heat transfer. Under convective heat transfer in a fluid with varying properties and in boiling, heat transfer coefficient may substantially depend on $\dot{q}_W$ and ΔT. In these cases an increase of heat flux may give rise to hazardous phenomena such as burnout (transition heat flux) and deterioration of turbulent heat transfer in tubes. If the $\dot{q}_W$ (ΔT) is nonlinear, it appears inappropriate to represent it in terms of the coefficient α when analyzing, for example, boiling stability.

An overall heat transfer coefficient

$$U = \dot{q}_W / (T_{f1} - T_{f2}) = \dot{q}_W / \nabla T_f,$$

where T_{f1} and T_{f2} are the temperatures of the heating and heated liquids, is used in calculations of heat transfer between two fluids through the separating wall. The U values for the most commonly used wall configurations are determined by the formulas

$$U = 1 / \left[\frac{1}{\alpha_1} + \sum_{i-1}^{n} \frac{S_i}{\lambda_i} + \frac{1}{\alpha_2} \right]$$

for a plane multilayer wall,

$$U = 1 / \left[\frac{1}{\alpha_1} \frac{D}{D_1} + \sum_{i-1}^{n} \frac{D}{2\lambda_i} \ln \frac{D_{i+1}}{D_i} + \frac{1}{\alpha_2} \frac{D}{D_2} \right]$$

for a cylindrical multilayer wall, and

$$U = 1 / \left[\frac{1}{\alpha_1} \frac{D^2}{D_1^2} + \sum_{i-1}^{n} \frac{(D_{i+1} - D_i)}{2\lambda_i} \frac{D^2}{D_i D_{i+1}} + \frac{1}{\alpha_2} \frac{D}{D_2} \right]$$

for a spherical multilayer wall.

Here D_1 and D_2 are the internal and external diameters of the wall, D the reference diameter by which a reference heat transfer surface is determined, S_i, D_i, D_{i+1} and λ_i are the thickness, internal and external diameters, and the thermal conductivity of the ith layer. The first and the third terms in brackets are said to be thermal resistances of heat transfer. In order to lower them the walls are finned and various methods of heat transfer augmentation are used. The second term in brackets is said to be the thermal resistance of the wall, which may greatly increase as a result of wall contamination, such as scale and ash build-up, or poor heat transfer between the wall layers. The values of α and U for a small element of heat transfer surface are called local ones. If they do not vary greatly then, in practical calculations of heat transfer on finite-size surfaces, we use the mean values of the coefficients and the heat transfer equation

$$Q = \overline{q_w} A = \overline{U} A \overline{\Delta T_{Mf}} \text{ or } \overline{\alpha} A \overline{\Delta T_{M_w}},$$

where A is the reference heat transfer surface, and $\overline{\Delta T_M}$ (ofen mean logarithmic $\overline{\Delta T_{LM}}$) temperature drop.

Table: Approximate values of heat transfer coefficient

Conditions of heat transfer	W/(m²K)
Gases in free convection	5-37
Water in free convection	100-1200
Oil under free convection	50-350
Gas flow in tubes and between tubes	10-350
Water flowing in tubes	500-1200
Oil flowing in tubes	300-1700
Molten metals flowing in tubes	2000-45000
Water nucleate boiling	2000-45000
Water film boiling	100-300
Film-type condensation of water vapour	4000-17000
Dropsize condensation of water vapour	30000-140000
Condensation of organic liquids	500-200

Heat Transfer Physics

Heat transfer physics describes the kinetics of storage, transport, and transformation of microscale energy carriers (phonon, electron, fluid particle, and photon). Sensible heat is stored in the thermal motion of atoms in various phases of matter. The atomic energy states and their populations are described by the classical and the quantum statistical mechanics (partition function and combinatoric energy distribution probabilities). Transport of thermal energy by the microscale carriers is based on their particle, quasi-particle, and wave descriptions; their diffusion, flow, and propagation; and their scattering and transformation encountered as they travel. The mechanisms of energy transitions among these energy carriers, and their rates (kinetics), are governed by the match of their energies, their interaction probabilities, and the various hindering-mechanism rate (kinetics) limits. Conservation of energy describes the interplay among energy storage, transport, and conversion, from the atomic to the continuum scales.

Heat is thermal energy associated with temperature-dependent motion of particles. The macroscopic energy equation for infinitesimal volume used in heat transfer analysis is

$$\nabla \cdot \mathbf{q} = -\rho c_p \frac{\partial T}{\partial t} + \sum_{i,j} \dot{s}_{i-j},$$

where q is heat flux vector, $-\rho c_p(\partial T/\partial t)$ is temporal change of internal energy (ρ is density, c_p is specific heat capacity at constant pressure, T is temperature and t is time), and $\dot{s}$ is the energy conversion to and from thermal energy (i and j are for principal energy carriers). So, the terms represent energy transport, storage and transformation. Heat flux vector q is composed of three macroscopic fundamental modes, which are conduction ($q_k = -k\nabla T$, k: thermal conductivity), convection ($q_u = \rho c_p uT$, u: velocity), and radiation ($q_r = 2\pi\int_0^\infty \int_0^\pi s\, I_{ph,\omega} \sin\theta d\theta d\omega$, ω: angular frequency, θ: polar angle, $I_{ph,\omega}$: spectral, directional radiation intensity, s: unit vector), i.e., $q = q_k + q_u + q_r$.

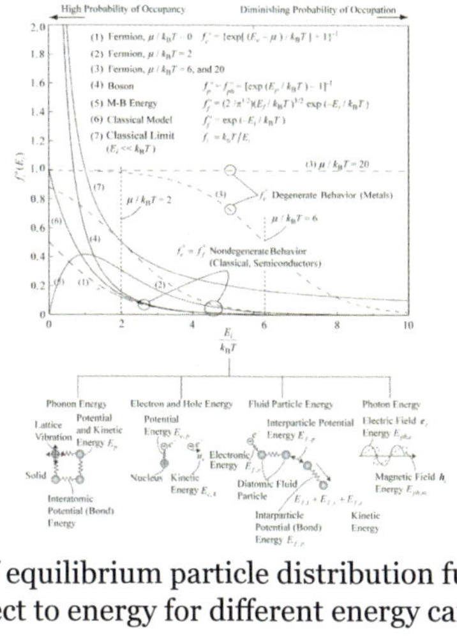

Variation of equilibrium particle distribution function with respect to energy for different energy carriers

Once states and kinetics of the energy conversion and thermophysical properties are known, the fate of heat transfer is described by the above equation. These atomic-level mechanisms and kinetics are addressed in heat transfer physics. The microscopic thermal energy is stored, transported, and transformed by the principal energy carriers: phonons (p), electrons (e), fluid particles (f), and photons (ph).

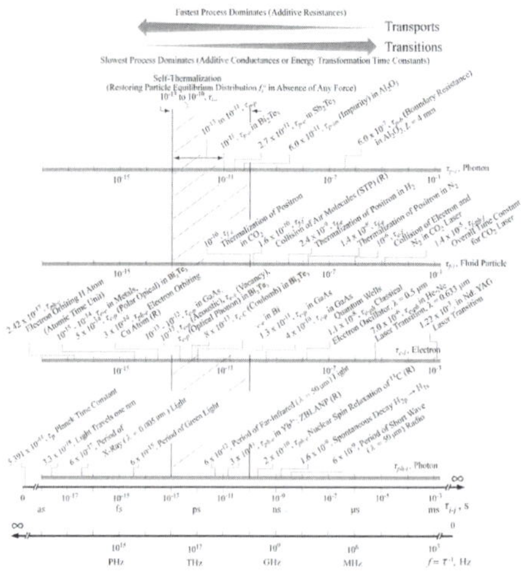

Kinetics of atomic-level energy transport and transition interaction

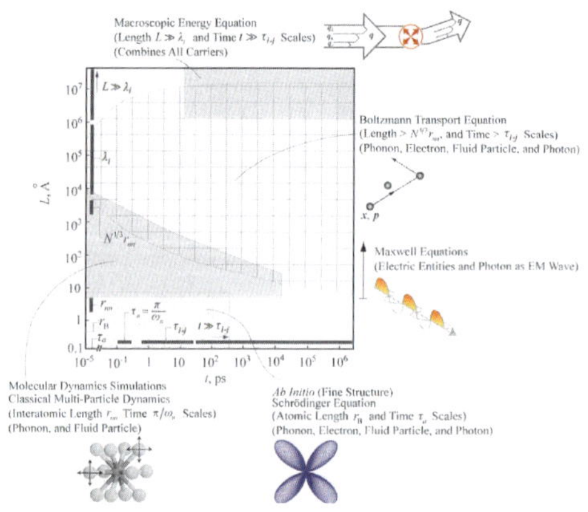

Length-time scale regimes for ab initio, MD, Boltzmann transport, and macroscopic treatments of heat transfer

Length and Time Scales

Thermophysical properties of matter and the kinetics of interaction and energy exchange among the principal carriers are based on the atomic-level configuration and interaction. Transport properties such as thermal conductivity are calculated from these atomic-level properties using classical and quantum physics. Quantum states of principal carriers (e.g. momentum, energy) are derived from the Schrödinger equation (called first principle or *ab initio*) and the interaction rates (for kinetics) are calculated using the quantum states and the quantum perturbation theory (formulated as the Fermi golden rule). Variety of *ab initio* solvers (software) exist (e.g., ABINIT, CASTEP, Gaussian, Q-Chem, Quantum ESPRESSO, SIESTA, VASP, WIEN2k). Electrons in the inner shells (core) are not involved in heat transfer, and calculations are greatly reduced by proper approximations about the inner-shells electrons.

The quantum treatments, including equilibrium and nonequilibrium *ab initio* molecular dynamics (MD), involving larger lengths and times are limited by the computation resources, so various alternate treatments with simplifying assumptions have been used and kinetics. In classical (Newtonian) MD, the motion of atom or molecule (particles) is based on the empirical or effective interaction potentials, which in turn can be based on curve-fit of *ab initio* calculations or curve-fit to thermophysical properties. From the ensembles of simulated particles, static or dynamics thermal properties or scattering rates are derived.

At yet larger length scales (mesoscale, involving many mean free paths), the Boltzmann transport equation (BTE) which is based on the classical Hamiltonian-statistical mechanics is applied. BTE considers particle states in terms of position and momentum vectors (x, p) and this is represented as the sate occupation probability. The occupation has equilibrium distributions (the known boson, fermion, and Maxwell–Boltzmann particles) and transport of energy (heat) is due to nonequilibrium (cause by a driving force or potential). Central to the transport is the role of scattering which turn the distribution toward equilibrium). The scattering is presented by the relations time or the mean free path. The relaxation time (or its inverse which is the interaction rate) is found from other calculations (*ab initio* or MD) or empirically. BTE can be numerically solved with Monte Carlo method, etc.

Depending on the length and time scale the proper level of treatment (*ab initio*, MD, or BTE). Heat transfer physics analyses may involve multiple scales (e.g., BTE using interaction rate from *ab initio* or classical MD) with states and kinetic related to thermal energy storage, transport and transformation.

So, heat transfer physics covers the four principal energy carries and their kinetics from classical and quantum mechanical perspectives. This enables multiscale (*ab initio*, MD, BTE and macroscale) analyses, including low-dimensionality and size effects.

Phonon

Phonon (quantized lattice vibration wave) is a central thermal energy carrier contributing to heat capacity (sensible heat storage) and conductive heat transfer in condensed phase, and plays very important role in thermal energy conversion. Its transport properties are represented by the phonon conductivity tensor K_p (W/m-K, from the Fourier law $q_{k,p} = -K_p \cdot \nabla T$) for bulk materials, and the phonon boundary resistance $AR_{p,b}$ [K/(W/m^2)] for solid interfaces, where A is the interface area. The phonon specific heat capacity $c_{v,p}$ (J/kg-K) includes the quantum effect. The thermal energy conversion rate involving phonon is included in $\dot{s}_{i\text{-}j}$. Heat transfer physics describes and predicts, $c_{v,p}$, K_p, $R_{p,b}$ (or conductance $G_{p,b}$) and $\dot{s}_{i\text{-}j}$, based on atomic-level properties.

For an equilibrium potential $\langle\varphi\rangle_o$ of a system with N atoms, the total potential $\langle\varphi\rangle$ is found by a Taylor series expansion at the equilibrium and this can be approximated by the second derivatives (the harmonic approximation) as

$$\langle\varphi\rangle = \langle\varphi\rangle_o + \sum_i \sum_\alpha \frac{\partial\langle\varphi\rangle}{\partial d_{i\alpha}}\Big|_o d_{i\alpha} + \frac{1}{2}\sum_{i,j}\sum_{\alpha,\beta} \frac{\partial^2\langle\varphi\rangle}{\partial d_{i\alpha}\partial d_{j\beta}}\Big|_o d_{i\alpha}d_{j\beta} + \frac{1}{6}\sum_{i,j,k}\sum_{\alpha,\beta,\gamma} \frac{\partial^3\langle\varphi\rangle}{\partial d_{i\alpha}\partial d_{j\beta}\partial d_{k\gamma}}\Big|_o d_{i\alpha}d_{j\beta}d_{k\gamma} + \ldots +$$

$$\approx \langle\varphi\rangle_o + \frac{1}{2}\sum_{i,j}\sum_{\alpha,\beta}\Gamma_{\alpha\beta} d_{i\alpha} d_{j\beta},$$

where d_i is the displacement vector of atom i, and Γ is the spring (or force) constant as the second-order derivatives of the potential. The equation of motion for the lattice vibration in terms of the displacement of atoms [$d(jl,t)$: displacement vector of the j-th atom in the l-th unit cell at time t] is

$$m_j \frac{d^2\mathbf{d}(jl,t)}{dt^2} = -\sum_{j'l'} \Gamma\begin{pmatrix} j\,j' \\ l\,l' \end{pmatrix} \cdot \mathbf{d}(j'l',T),$$

where m is the atomic mass and Γ is the force constant tensor. The atomic displacement is the summation over the normal modes [s_α: unit vector of mode α, ω_p: angular frequency of wave, and κ_p: wave vector]. Using this plane-wave displacement, the equation of motion becomes the eigenvalue equation

$$\mathrm{M}\,\omega_p^2(\kappa_p,\alpha)\mathrm{s}_\alpha(\kappa_p) = \mathrm{D}(\kappa_p)\mathbf{s}_\alpha(\kappa_p),$$

where M is the diagonal mass matrix and D is the harmonic dynamical matrix. Solving this eigenvalue equation gives the relation between the angular frequency ω_p and the wave vector κ_p, and this relation is called the phonon dispersion relation. Thus, the phonon dispersion relation is determined by matrices M and D, which depend on the atomic structure and the strength of interaction among constituent atoms (the stronger the interaction and the lighter the atoms, the higher is the phonon frequency and the larger is the slope $d\omega_p/d\kappa_p$). The Hamiltonian of phonon system with the harmonic approximation is

$$H_p = \sum_x \frac{1}{2m}\mathbf{p}^2(\mathbf{x}) + \frac{1}{2}\sum_{\mathbf{x},\mathbf{x}'} \mathbf{d}_i(\mathbf{x}) D_{ij}(\mathbf{x}-\mathbf{x}')\mathbf{d}_j(\mathbf{x}'),$$

where D_{ij} is the dynamical matrix element between atoms i and j, and d_i (d_j) is the displacement of i (j) atom, and p is momentum. From this and the solution to dispersion relation, the phonon annihilation operator for the quantum treatment is defined as

$$b_{\kappa,\alpha} = \frac{1}{N^{1/2}}\sum_{\kappa_p,\alpha} e^{-i(\kappa_p\cdot\mathbf{x})}\mathbf{s}_\alpha(\kappa_p)\cdot[(\frac{m\omega_{p,\alpha}}{2\hbar})^{1/2}\mathbf{d}(\mathrm{x}) + i(\frac{1}{2\hbar m\omega_{p,\alpha}})^{1/2}\,\mathrm{p}(\mathbf{x})],$$

where N is the number of normal modes divided by α and $\hbar$ is the reduced Planck constant. The creation operator is the adjoint of the annihilation operator,

$$b^{\dagger}_{\kappa,\alpha} = \frac{1}{N^{1/2}} \sum_{\kappa_p,\alpha} e^{i(\hat{\mathbf{e}}_p \cdot \mathbf{x})} \mathbf{s}_{\alpha}(\kappa_p) \cdot [(\frac{m\omega_{p,\alpha}}{2\hbar})^{1/2} \mathbf{d}(\mathbf{x}) - i(\frac{1}{2\hbar m\omega_{p,\alpha}})^{1/2} \mathbf{p}(\mathbf{x})].$$

The Hamiltonian in terms of $b_{\kappa,\alpha}{}^{\dagger}$ and $b_{\kappa,\alpha}$ is $H_p = \sum_{\kappa,\alpha} \hbar\omega_{p,\alpha}[b_{\kappa,\alpha}{}^{\dagger} b_{\kappa,\alpha} + 1/2]$ and $b_{\kappa,\alpha}{}^{\dagger} b_{\kappa,\alpha}$ is the phonon number operator. The energy of quantum-harmonic oscillator is $E_p = \sum_{\kappa,\alpha} [f_p(\kappa,\alpha) + 1/2]\hbar\omega_{p,\alpha}(\kappa_p)$, and thus the quantum of phonon energy $\hbar\omega_p$.

The phonon dispersion relation gives all possible phonon modes within the Brillouin zone (zone within the primitive cell in reciprocal space), and the phonon density of states D_p (the number density of possible phonon modes). The phonon group velocity $u_{p,g}$ is the slope of the dispersion curve, $d\omega_p/d\kappa_p$. Since phonon is a boson particle, its occupancy follows the Bose–Einstein distribution $\{f_p{}^{o} = [\exp(\hbar\omega_p/k_B T)-1]^{-1}$, k_B: Boltzmann constant$\}$. Using the phonon density of states and this occupancy distribution, the phonon energy is $E_p(T) = \int D_p(\omega_p) f_p(\omega_p,T)\hbar\omega_p d\omega_p$, and the phonon density is $n_p(T) = \int D_p(\omega_p) f_p(\omega_p,T) d\omega_p$. Phonon heat capacity $c_{v,p}$ (in solid $c_{v,p} = c_{p,p}$, $c_{v,p}$: constant-volume heat capacity, $c_{p,p}$: constant-pressure heat capacity) is the temperature derivatives of phonon energy for the Debye model (linear dispersion model), is

$$c_{v,p} = \frac{dE_p}{dT}\Big|_v = \frac{9k_B}{m}\left(\frac{T}{T_D}\right)^3 n \int_0^{T_D/T} \frac{x^4 e^x}{(e^x-1)^2} dx \qquad \left(x = \frac{\hbar\omega}{k_B T}\right),$$

where T_D is the Debye temperature, m is atomic mass, and n is the atomic number density (number density of phonon modes for the crystal $3n$). This gives the Debye T^3 law at low temperature and Dulong-Petit law at high temperatures.

From the kinetic theory of gases, thermal conductivity of principal carrier i (p, e, f and ph) is

$$k_i = \frac{1}{3} n_i c_{v,i} u_i \lambda_i,$$

where n_i is the carrier density and the heat capacity is per carrier, u_i is the carrier speed and λ_i is the mean free path (distance traveled by carrier before an scattering event). Thus, the larger the carrier density, heat capacity and speed, and the less significant the scattering, the higher is the conductivity. For phonon λ_p represents the interaction (scattering) kinetics of phonons and is related to the scattering relaxation time T_p or rate ($= 1/T_p$) through $\lambda_p = u_p T_p$. Phonons interact with other phonons, and with electrons, boundaries, impurities, etc., and λ_p combines these interaction mechanisms through the Matthiessen rule. At low temperatures, scattering by boundaries is dominant and with increase in temperature the interaction rate with impurities, electron and other phonons become important, and finally the phonon-phonon scattering dominants for $T > 0.2T_D$. The interaction rates are reviewed in and includes quantum perturbation theory and MD.

A number of conductivity models are available with approximations regarding the dispersion and λ_p. Using the single-mode relaxation time approximation ($\partial f_p'/\partial t|_s = -f_p'/T_p$) and the gas kinetic theory, Callaway phonon (lattice) conductivity model as

$$k_{p,s} = \frac{1}{8\pi^3}\sum_{\alpha}\int c_{v,p}\tau_p(\mathbf{u}_{p,g}\cdot\mathbf{s})^2 d\kappa \qquad \textit{for component along } \mathbf{s},$$

$$k_p = \frac{1}{6\pi^3}\sum_{\alpha}\int c_{v,p}\tau_p u_{p,g}^2\kappa^2 d\kappa \qquad \textit{for isotropic conductivity.}$$

With the Debye model (a single group velocity $u_{p,g}$, and a specific heat capacity calculated above), this becomes

$$k_p = (48\pi^2)^{1/3}\frac{k_B^3 T^3}{a h_P^2 T_D}\int_0^{T/T_D}\tau_p\frac{x^4 e^x}{(e^x-1)^2}dx,$$

where a is the lattice constant $a = n^{-1/3}$ for a cubic lattice, and n is the atomic number density. Slack phonon conductivity model mainly considering acoustic phonon scattering (three-phonon interaction) is given as

$$k_p = k_{p,S} = \frac{3.1\times10^{12}\langle M\rangle V_a^{1/3}T_{D,\infty}^3}{T\langle\gamma_G^2\rangle N_o^{2/3}} \qquad \textit{high temperatures } (T > 0.2T_D, \textit{ phonon-phonon scattering only}),$$

where $\langle M\rangle$ is the mean atomic weight of the atoms in the primitive cell, $V_a = 1/n$ is the average volume per atom, $T_{D,\infty}$ is the high-temperature Debye temperature, T is the temperature, N_o is the number of atoms in the primitive cell, and $\langle\gamma_G^2\rangle$ is the mode-averaged square of the Grüneisen constant or parameter at high temperatures. This model is widely tested with pure nonmetallic crystals, and the overall agreement is good, even for complex crystals.

Based on the kinetics and atomic structure consideration, a material with high crystalline and strong interactions, composed of light atoms (such as diamond and graphene) is expected to have large phonon conductivity. Solids with more than one atom in the smallest unit cell representing the lattice have two types of phonons, i.e., acoustic and optical. (Acoustic phonons are in-phase movements of atoms about their equilibrium positions, while optical phonons are out-of-phase movement of adjacent atoms in the lattice.) Optical phonons have higher energies (frequencies), but make smaller contribution to conduction heat transfer, because of their smaller group velocity and occupancy.

Phonon transport across hetero-structure boundaries (represented with $R_{p,b}$, phonon boundary resistance) according to the boundary scattering approximations are modeled as acoustic and diffuse mismatch models. Larger phonon transmission (small $R_{p,b}$) occurs at boundaries where material pairs have similar phonon properties (u_p, D_p, etc.), and in contract large $R_{p,b}$ occurs when some material is softer (lower cut-off phonon frequency) than the other.

Electron

Quantum electron energy states for electron are found using the electron quantum Hamiltonian, which is generally composed of kinetic ($-\hbar^2\nabla^2/2m_e$) and potential energy terms (φ_e). Atomic orbital, a mathematical function describing the wave-like behavior of either an electron or a pair of electrons in an atom, can be found from the Schrödinger equation with this electron Hamiltonian. Hydrogen-like atoms (a nucleus and an electron) allow for closed-form solution to Schrödinger equation with the electrostatic potential (the Coulomb law). The Schrödinger equation of atoms or atomic ions with more than one electron has not been solved analytically, because of the Coulomb interactions among electrons. Thus, numerical techniques are used, and an electron configuration is approximated as product of simpler hydrogen-like atomic orbitals (isolate electron orbitals). Molecules with multiple atoms (nuclei and their electrons) have molecular orbital (MO, a mathematical function for the wave-like behavior of an electron in a molecule), and are obtained from simplified solution techniques such as linear combination of atomic orbitals (LCAO). The molecular orbital is used to predict chemical and physical properties, and the difference between highest occupied molecular orbital (HOMO) and the lowest unoccupied molecular orbital (LUMO) is a measure of excitability of the molecules.

In a crystal structure of metallic solids, the free electron model (zero potential, $\varphi_e = 0$) for the behavior of valence electrons is used. However, in a periodic lattice (crystal), there is periodic crystal potential, so the electron Hamiltonian becomes

$$H_e = -\frac{\hbar^2}{2m_e}\nabla^2 + \varphi_c(\mathbf{x}),$$

where m_e is the electron mass, and the periodic potential is expressed as $\varphi_c(x) = \sum_g \varphi_g \exp[i(g\cdot x)]$ (g: reciprocal lattice vector). The time-independent Schrödinger equation with this Hamiltonian is given as (the eigenvalue equation)

$$H_e \psi_{e,\mathbf{x}}(\mathbf{x}) = E_e(\kappa_e)\psi_{e,\mathbf{x}}(\mathbf{x}),$$

where the eigenfunction $\psi_{e,\kappa}$ is the electron wave function, and eigenvalue $E_e(\kappa_e)$, is the electron energy (κ_e: electron wavevector). The relation between wavevector, κ_e and energy E_e provides the electronic band structure. In practice, a lattice as many-body systems includes interactions between electrons and nuclei in potential, but this calculation can be too intricate. Thus, many approximate techniques have been suggested and one of them is density functional theory (DFT), uses functionals of the spatially dependent electron density instead of full interactions. DFT is widely used in *ab initio* software (ABINIT, CASTEP, Quantum ESPRESSO, SIESTA, VASP, WIEN2k, etc.). The electron specific heat is based on the energy states and occupancy distribution (the Fermi–Dirac statistics). In general, the heat capacity of electron is small except at very high temperature when they are in thermal equilibrium with phonons (lattice). Electrons contribute to heat conduction (in addition to charge carrying) in solid, especially in metals. Thermal conductivity tensor in solid is the sum of electric and phonon thermal conductivity tensors $K = K_e + K_p$.

Electrons are affected by two thermodynamic forces [from the charge, $\nabla(E_F/e_c)$ where E_F is the Fermi level and e_c is the electron charge and temperature gradient, $\nabla(1/T)$] because they carry both charge and thermal energy, and thus electric current j_e and heat flow q are described with the thermoelectric tensors (A_{ee}, A_{et}, A_{te}, and A_{tt}) from the Onsager reciprocal relations as

$$\mathbf{j}_e = \mathbf{A}_{ee} \cdot \nabla \frac{E_F}{e_c} + \mathbf{A}_{et} \cdot \nabla \frac{1}{T}, \text{ and}$$

$$\mathbf{q} = \mathbf{A}_{te} \cdot \nabla \frac{E_F}{e_c} + \mathbf{A}_{tt} \cdot \nabla \frac{1}{T}.$$

Converting these equations to have j_e equation in terms of electric field e_e and ∇T and q equation with j_e and ∇T, (using scalar coefficients for isotropic transport, α_{ee}, α_{et}, α_{te}, and α_{tt} instead of A_{ee}, A_{et}, A_{te}, and A_{tt})

$$\mathbf{j}_e = \alpha_{ee}\mathbf{e}_e - \frac{\alpha_{et}}{T^2}\nabla T \qquad (\mathbf{e}_e = \alpha_{ee}^{-1}\mathbf{j}_e + \frac{\alpha_{ee}^{-1}\alpha_{et}}{T^2}\nabla T),$$

$$\mathbf{q} = \alpha_{te}\alpha_{ee}^{-1}\mathbf{j}_e - \frac{\alpha_{tt} - \alpha_{te}\alpha_{ee}^{-1}\alpha_{et}}{T^2}\nabla T.$$

Electrical conductivity/resistivity σ_e (Ω^{-1}m^{-1})/ ρ_e (Ω-m), electric thermal conductivity k_e (W/m-K) and the Seebeck/Peltier coefficients α_S (V/K)/α_P (V) are defined as,

$$\sigma_e = \frac{1}{\rho_e} = \alpha_{ee},\ k_e = \frac{\alpha_{tt} - \alpha_{te}\alpha_{ee}^{-1}\alpha_{et}}{T^2}, and\ \alpha_S = \frac{\alpha_{et}\alpha_{ee}^{-1}}{T^2} \quad (\alpha_S = \alpha_P T).$$

Various carriers (electrons, magnons, phonons, and polarons) and their interactions substantially affect the Seebeck coefficient. The Seebeck coefficient can be decomposed with two contributions, $\alpha_S = \alpha_{S,pres} + \alpha_{S,trans}$, where $\alpha_{S,pres}$ is the sum of contributions to the carrier-induced entropy change, i.e., $\alpha_{S,pres} = \alpha_{S,mix} + \alpha_{S,spin} + \alpha_{S,vib}$ ($\alpha_{S,mix}$: entropy-of-mixing, $\alpha_{S,spin}$: spin entropy, and $\alpha_{S,vib}$: vibrational entropy). The other contribution $\alpha_{S,trans}$ is the net energy transferred in moving a carrier divided by qT (q: carrier charge). The electron's contributions to the Seebeck coefficient are mostly in $\alpha_{S,pres}$. The $\alpha_{S,mix}$ is usually dominant in lightly doped semiconductors. The change of the entropy-of-mixing upon adding an electron to a system is the so-called Heikes formula

$$\alpha_{S,mix} = \frac{1}{q}\frac{\partial S_{mix}}{\partial N} = \frac{k_B}{q} ln(\frac{1-f_e^o}{f_e^o}),$$

where $f_e^o = N/N_a$ is the ratio of electrons to sites (carrier concentration). Using the chemical potential (μ), the thermal energy (k_BT) and the Fermi function, above equation can be expressed in an alternative form, $\alpha_{S,mix} = (k_B/q)[(E_e - \mu)/(k_B T)]$. Extending the Seebeck effect to spins, a ferromagnetic alloy can be a good example. The contribution to the Seebeck coefficient that results from electrons' presence altering the systems spin entropy is given by $\alpha_{S,spin} = \Delta S_{spin}/q = (k_B/q)\ln[(2s + 1)/(2s_0 + 1)]$, where s_0 and s are net spins of the magnetic site in the absence and presence of the carrier, respectively. Many vibrational effects with electrons also contribute to the Seebeck coefficient. The softening of the vibrational frequencies produces a change of the vibrational entropy is one of examples. The vibrational entropy is the negative derivative of the free energy, i.e.,

$$S_{vib} = -\frac{\partial F_{mix}}{\partial T} = 3Nk_B T\int_0^{\omega}\{\frac{\hbar\omega}{2k_B T}coth(\frac{\hbar\omega}{2k_B T}) - ln[2sinh(\frac{\hbar\omega}{2k_B T})]\}D_p(\omega)d\omega,$$

where $D_p(\omega)$ is the phonon density-of-states for the structure. For the high-temperature limit and series expansions of the hyperbolic functions, the above is simplified as $\alpha_{S,vib} = (\Delta S_{vib}/q) = (k_B/q)\Sigma_i(-\Delta\omega_i/\omega_i)$.

The Seebeck coefficient derived in the above Onsager formulation is the mixing component $\alpha_{S,mix}$, which dominates in most semiconductors. The vibrational component in high-band gap materials such as $B_{13}C_2$ is very important. Considering the microscopic transport (transport is a results of nonequilibrium),

$$\mathbf{j}_e = -\frac{e_c}{\hbar^3}\sum_p \mathbf{u}_e f_e' = -\frac{e_c}{\hbar^3 k_B T}\sum_p \mathbf{u}_e \tau_e(-\frac{\partial f_e^o}{\partial E_e})(\mathbf{u}_e \cdot \mathbf{F}_{te}),$$

$$\mathbf{q} = \frac{1}{\hbar^3}\sum_p (E_e - E_F)\mathbf{u}_e f_e' = \frac{1}{\hbar^3 k_B T}\sum_p \mathbf{u}_e \tau_e(-\frac{\partial f_e^o}{\partial E_e})(E_e - E_F)(\mathbf{u}_e \cdot \mathbf{F}_{te}),$$

where u_e is the electron velocity vector, f_e' (f_e^o) is the electron nonequilibrium (equilibrium) distribution, τ_e is the electron scattering time, E_e is the electron energy, and F_{te} is the electric and thermal forces from $\nabla(E_F/e_c)$ and $\nabla(1/T)$. Relating the thermoelectric coefficients to the microscopic transport equations for j_e and q, the thermal, electric, and thermoelectric properties are calculated. Thus, k_e increases with the electrical conductivity σe and temperature T, as the Wiedemann–Franz law presents $[k_e/(\sigma_e T_e) = (1/3)(\pi k_B/e_c)^2 = 2.44\times10^{-8}\ \text{W-}\Omega/\text{K}^2]$. Electron transport (represented as σ_e) is a function of carrier density $n_{e,c}$ and electron mobility μ_e ($\sigma_e = e_c n_{e,c}\mu_e$). μ_e is determined by electron scattering rates $\dot{\gamma}_e$ (or relaxation time, $\tau_e = 1/\dot{\gamma}_e$) in various interaction mechanisms including interaction with other electrons, phonons, impurities and boundaries.

Electrons interact with other principal energy carriers. Electrons accelerated by an electric field are relaxed through the energy conversion to phonon (in semiconductors, mostly optical phonon), which is called Joule heating. Energy conversion between electric potential and phonon energy is considered in thermoelectrics such as Peltier cooling and thermoelectric generator. Also, study of interaction with photons is central in optoelectronic applications (i.e. light-emitting diode, solar photovoltaic cells, etc.). Interaction rates or energy conversion rates can be evaluated using the Fermi golden rule (from the perturbation theory) with *ab initio* approach.

Fluid Particle

Fluid particle is the smallest unit (atoms or molecules) in the fluid phase (gas, liquid or plasma) without breaking any chemical bond. Energy of fluid particle is divided into potential, electronic, translational, vibrational, and rotational energies. The heat (thermal) energy storage in fluid particle is through the temperature-dependent particle motion (translational, vibrational, and rotational energies). The electronic energy is included only if temperature is high enough to ionize or dissociate the fluid particles or to include other electronic transitions. These quantum energy states of the fluid particles are found using their respective quantum Hamiltonian. These are $H_{f,t} = -(\hbar^2/2m)\nabla^2$, $H_{f,v} = -(\hbar^2/2m)\nabla^2 + \Gamma x^2/2$ and $H_{f,r} = -(\hbar^2/2I_f)\nabla^2$ for translational, vibrational and rotational modes. (Γ: spring constant, I_f: the moment of inertia for the molecule). From the Hamiltonian, the quantized fluid particle energy state E_f and partition functions Z_f [with the Maxwell–Boltzmann (MB) occupancy distribution] are found as

$$\textit{translational} \qquad E_{f,t,n} = \frac{\pi^2\hbar^2}{2m}\left(\frac{n_x^2}{L^2}+\frac{n_y^2}{L^2}+\frac{n_z^2}{L^2}\right) \text{ and } Z_{f,t}\sum_{i=0}^{\infty} g_{f,t,i} exp\left(-\frac{E_{f,t,i}}{k_B T}\right) = V\left(\frac{mk_B T}{2\pi\hbar^2}\right)^{3/2},$$

$$\textit{vibrational} \qquad E_{f,v,l} = \hbar\omega_{f,v}\left(1+\frac{1}{2}\right) \text{ and } Z_{f,v}\sum_{j=0}^{\infty} exp\left[-\left(l+\frac{1}{2}\right)\frac{\hbar\omega_{f,v}}{k_B T}\right] = \frac{exp(-T_{f,v}/2T)}{1-exp(-T_{f,v}/T)},$$

$$\textit{rotational} \qquad E_{f,r,j} = \frac{\hbar^2}{2I_f} \text{ and } Z_{f,r}\sum_{j=0}^{\infty}(2j+1)exp\left[-\frac{-\hbar^2 j(j+1)}{2I_f k_B T}\right] \approx \frac{T}{T_{f,r}}\left(1+\frac{T_{f,r}}{3T}+\frac{T_{f,r}^2}{15T^2}+\ldots\right),$$

$$\textit{total} \qquad E_f = \sum_i E_{f,i} = E_{f,t} + E_{f,v} + E_{f,r} + \ldots \text{ and } Z_f = \prod_i Z_{f,i} = Z_{f,t} Z_{f,v} Z_{f,r}\ldots.$$

Here, g_f is the degeneracy, n, l, and j are the transitional, vibrational and rotational quantum numbers, $T_{f,v}$ is the characteristic temperature for vibration ($= \hbar\omega_{f,v}/k_B$, : vibration frequency), and $T_{f,r}$ is the rotational temperature [$= \hbar^2/(2I_f k_B)$]. The average specific internal energy is related to the partition function through Z_f, $e_f = (k_B T^2/m)(\partial \ln Z_f/\partial T)|_{N,V}$.

With the energy states and the partition function, the fluid particle specific heat capacity $c_{v,f}$ is the summation of contribution from various kinetic energies (for non-ideal gas the potential energy is also added). Because the total degrees of freedom in molecules is determined by the atomic configuration, $c_{v,f}$ has different formulas depending on the configuration,

$$\textit{monatomic ideal gas} \qquad c_{v,f} = \frac{\partial e_f}{\partial T}\Big|_V = \frac{3R_g}{2M},$$

$$\textit{diatomic ideal gas} \qquad c_{v,f} = \frac{R_g}{M}\left\{\frac{3}{2}+\left(\frac{T_{f,v}}{T}\right)^2\frac{exp(T_{f,v,i}/T)}{[exp(T_{f,v,i}/T)-1]^2}+1+\frac{2}{15}\left(\frac{T_{f,v}}{T}\right)^2\right\},$$

$$\textit{nonlinear, polyatomic ideal gas} \qquad c_{v,f} = \frac{R_g}{M}\left\{3+\sum_{j=1}^{3N_o-6}\left(\frac{T_{f,v}}{T}\right)^2\frac{exp(T_{f,v,i}/T)}{[exp(T_{f,v,i}/T)-1]^2}\right\}.$$

where R_g is the gas constant ($= N_A k_B$, N_A: the Avogadro constant) and M is the molecular mass (kg/kmole). (For the polyatomic ideal gas, N_o is the number of atoms in a molecule.) In gas, constant-pressure specific heat capacity $c_{p,f}$ has a larger value and the difference depends on the temperature T, volumetric thermal expansion coefficient β and the isothermal compressibility κ [$c_{p,f} - c_{v,f} = T\beta^2/(\rho_f \kappa)$, ρ_f: the fluid density]. For dense fluids that the interactions between the particles (the van der Waals interaction) should be included, and $c_{v,f}$ and $c_{p,f}$ would change accordingly. The net motion of particles (under gravity or external pressure) gives rise to the convection heat flux $q_u = \rho_f c_{p,f} u_f T$. Conduction heat flux q_k for ideal gas is derived with the gas kinetic theory or the Boltzmann transport equations, and the thermal conductivity is

$$k_f = \frac{1}{3} n_f c_{p,f} \langle u_f^2 \rangle \tau_{f-f},$$

where $\langle u_f^2 \rangle^{1/2}$ is the RMS (root mean square) thermal velocity ($3k_B T/m$ from the MB distribution function, m: atomic mass) and $\tau_{f\text{-}f}$ is the relaxation time (or intercollision time period) [$(2^{1/2}\pi d^2 n_f \langle u_f \rangle)^{-1}$ from the gas kinetic theory, $\langle u_f \rangle$: average thermal speed $(8k_B T/\pi m)^{1/2}$, d: the collision diameter of fluid particle (atom or molecule), n_f: fluid number density].

k_f is also calculated using molecular dynamics (MD), which simulates physical movements of the fluid particles with the Newton equations of motion (classical) and force field (from *ab initio* or empirical properties). For calculation of k_f, the equilibrium MD with Green–Kubo relations, which express the transport coefficients in terms of integrals of time correlation functions (considering fluctuation), or nonequilibrium MD (prescribing heat flux or temperature difference in simulated system) are generally employed.

Fluid particles can interact with other principal particles. Vibrational or rotational modes, which have relatively high energy, are excited or decay through the interaction with photons. Gas lasers employ the interaction kinetics between fluid particles and photons, and laser cooling has been also considered in CO_2 gas laser. Also, fluid particles can be adsorbed on solid surfaces (physisorption and chemisorption), and the frustrated vibrational modes in adsorbates (fluid particles) is decayed by creating e^--h^+ pairs or phonons. These interaction rates are also calculated through *ab initio* calculation on fluid particle and the Fermi golden rule.

Photon

Photon is the quanta of electromagnetic (EM) radiation and energy carrier for radiation heat transfer. The EM wave is governed by the classical Maxwell equations, and the quantization of EM wave is used for phenomena such as the blackbody radiation (in particular to explain the ultraviolet catastrophe). The quanta EM wave (photon) energy of angular frequency ω_{ph} is $E_{ph} = \hbar\omega_{ph}$, and follows the Bose–Einstein distribution function (f_{ph}). The photon Hamiltonian for the quantized radiation field (second quantization) is

$$H_{ph} = \frac{1}{2}\int (\epsilon_o \mathbf{e}_e^2 + \mu_o^{-1}\mathbf{b}_e^2)dV = \sum_\alpha \hbar\omega_{ph,\alpha}(c_\alpha^\dagger c_\alpha + \frac{1}{2}),$$

where e_e and b_e are the electric and magnetic fields of the EM radiation, ε_o and μ_o are the free-space permittivity and permeability, V is the interaction volume, $\omega_{ph,\alpha}$ is the photon angular frequency for the α mode and $c_\alpha^\dagger$ and c_α are the photon creation and annihilation operators. The vector potential a_e of EM fields ($e_e = -\partial a_e/\partial t$ and $b_e = \nabla \times a_e$) is

$$\mathbf{a}_e(\mathbf{x},t) = \sum_\alpha \left(\frac{\hbar}{2\epsilon_o \omega_{ph,\alpha} V}\right)^{1/2} \mathbf{s}_{ph,\alpha}(c_\alpha e^{i\hat{\mathbf{e}}_\alpha \cdot \mathbf{x}} + c_\alpha^\dagger e^{-i\hat{\mathbf{e}}_\alpha \cdot \mathbf{x}}),$$

where $s_{ph,\alpha}$ is the unit polarization vector, κ_α is the wave vector.

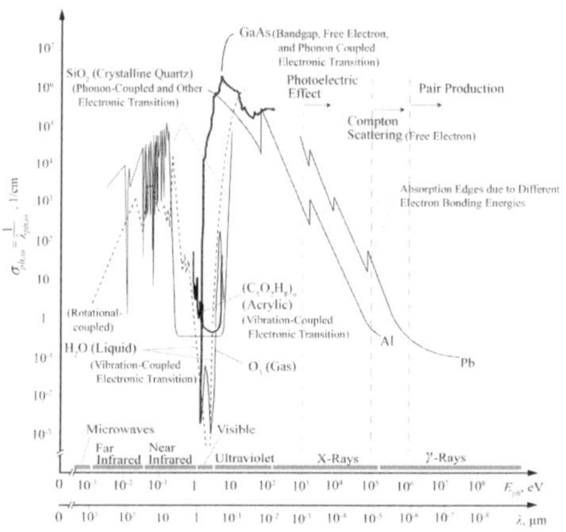

Spectral photon absorption coefficient for typical gas, liquid, and solid phases. For the solid phase, examples of polymer, oxide, semiconductor, and metals are given

Blackbody radiation among various types of photon emission employs the photon gas model with thermalized energy distribution without interphoton interaction. From the linear dispersion relation (i.e., dispersionless), phase and group speeds are equal ($u_{ph} = d\,\omega_{ph}/d\kappa = \omega_{ph}/\kappa$, u_{ph}: photon speed) and the Debye (used for dispersionless photon) density of states is $D_{ph,b,\omega}d\omega = \omega_{ph}^2 d\omega_{ph}/\pi^2 u_{ph}^3$. With $D_{ph,b,\omega}$ and equilibrium distribution f_{ph}, photon energy spectral distribution $dI_{b,\omega}$ or $dI_{b,\lambda}$ (λ_{ph}: wavelength) and total emissive power E_b are derived as

$$dI_{b,\omega} = \frac{D_{ph,b,\omega} f_{ph} u_{ph} d\omega_{ph}}{4\pi} = \frac{\hbar\omega_{ph}^3}{4\pi^3 u_{ph}^2}\frac{1}{e^{\hbar\omega_{ph}/k_B T} - 1}d\omega_{ph} \ or\ dI_{b,\lambda} = \frac{4\pi\hbar u_{ph}^2 d\lambda_{ph}}{\lambda_{ph}^5 (e^{2\pi\hbar u_{ph}/\lambda_{ph} k_B T} - 1)} \quad \text{(Planck law)}$$

$$E_b = \int_0^\infty dE_{b,\lambda} = \sigma_{SB} T^4, \ where\ \sigma_{SB} = \frac{\pi^2 k_B^4}{60\hbar^3 u_{ph}^2} \quad \text{(Stefan–Boltzmann law)}$$

Compared to blackbody radiation, laser emission has high directionality (small solid angle $\Delta\Omega$) and spectral purity (narrow bands $\Delta\omega$). Lasers range far-infrared to X-rays/γ-rays regimes based on the resonant transition (stimulated emission) between electronic energy states.

Near-field radiation from thermally excited dipoles and other electric/magnetic transitions is very effective within a short distance (order of wavelength) from emission sites.

The BTE for photon particle momentum $p_{ph} = \hbar\omega_{ph} s/u_{ph}$ along direction s experiencing absorption/emission $\dot{s}_{f,ph-e}$ $(= u_{ph}\sigma_{ph,\omega}[f_{ph}(\omega_{ph},T) - f_{ph}(s)]$, $\sigma_{ph,\omega}$: spectral absorption coefficient), and generation/removal $\dot{s}_{f,ph,i}$, is

$$\frac{\partial f_{ph}}{\partial t} + u_{ph}\mathbf{s}\cdot\nabla f_{ph} = \frac{\partial f_{ph}}{\partial t}|_s + u_{ph}\sigma_{ph,\omega}[f_{ph}(\omega_{ph},T) - f_{ph}(\mathbf{s})] + \dot{s}_{f,ph,i}.$$

In terms of radiation intensity ($I_{ph,\omega} = u_{ph} f_{ph}\hbar\omega_{ph} D_{ph,\omega}/4\pi$, $D_{ph,\omega}$: photon density of states), this is called the equation of radiative transfer (ERT)

$$\frac{\partial I_{ph,\omega}(\omega_{ph},\mathbf{s})}{u_{ph}\partial t} + \mathbf{s}\cdot\nabla I_{ph,\omega}(\omega_{ph},\mathbf{s}) = \frac{\partial I_{ph,\omega}(\omega_{ph},\mathbf{s})}{u_{ph}\partial t}|_s + \sigma_{ph,\omega}[I_{ph,\omega}(\omega_{ph},T) - I_{ph}(\omega_{ph},\mathbf{s})] + \dot{s}_{ph,i}.$$

The net radiative heat flux vector is $\mathbf{q}_r = \mathbf{q}_{ph} = \int_0^\infty \int_{4\pi} \mathbf{s} I_{ph,\omega} d\Omega d\omega.$

From the Einstein population rate equation, spectral absorption coefficient $\sigma_{ph,\omega}$ in ERT is,

$$\sigma_{ph,\omega} = \frac{\hbar\omega\dot{\gamma}_{ph,a} n_e}{u_{ph}},$$

where $\dot{\gamma}_{ph,a}$ is the interaction probability (absorption) rate or the Einstein coefficient B_{12} (J^{-1} m^3 s^{-1}), which gives the probability per unit time per unit spectral energy density of the radiation field (1: ground state, 2: excited state), and n_e is electron density (in ground state). This can be obtained using the transition dipole moment $\boldsymbol{\mu}_e$ with the FGR and relationship between Einstein coefficients. Averaging $\sigma_{ph,\omega}$ over ω gives the average photon absorption coefficient σ_{ph}.

For the case of optically thick medium of length L, i.e., $\sigma_{ph} L >> 1$, and using the gas kinetic theory, the photon conductivity k_{ph} is $16\sigma_{SB}T^3/3\sigma_{ph}$ (σ_{SB}: Stefan–Boltzmann constant, σ_{ph}: average photon absorption), and photon heat capacity $n_{ph}c_{v,ph}$ is $16\sigma_{SB}T^3/u_{ph}$.

Photons have the largest range of energy and central in a variety of energy conversions. Photons interact with electric and magnetic entities. For example, electric dipole which in turn are excited by optical phonons or fluid particle vibration, or transition dipole moments of electronic transitions. In heat transfer physics, the interaction kinetics of phonon is treated using the perturbation theory (the Fermi golden rule) and the interaction Hamiltonian. The photon-electron interaction is

$$H_{ph-e} = -\frac{e_c}{m_e}(a + a^\dagger)\mathbf{a}_e\cdot\mathbf{p}_e = -\left(\frac{\hbar\omega_{ph,\alpha}}{2\epsilon_o V}\right)^{1/2}(\mathbf{s}_{ph,\alpha}\cdot e_c\mathbf{x}_e)(a + a^\dagger)(ce^{i\kappa\cdot x} + c^\dagger e^{-i\kappa\cdot x}),$$

where p_e is the dipole moment vector and $a^\dagger$ and a are the creation and annihilation of internal motion of electron. Photons also participate in ternary interactions, e.g., phonon-assisted photon absorption/emission (transition of electron energy level). The vibrational mode in fluid particles can decay or become excited by emitting or absorbing photons. Examples are solid and molecular gas laser cooling.

Using *ab initio* calculations based on the first principles along with EM theory, various radiative properties such as dielectric function (electrical permittivity, $\varepsilon_{e,\omega}$), spectral absorption coefficient ($\sigma_{ph,\omega}$), and the complex refraction index (m_ω), are calculated for various interactions between photons and electric/magnetic entities in matter. For example, the imaginary part ($\varepsilon_{e,c,\omega}$) of complex dielectric function ($\varepsilon_{e,\omega} = \varepsilon_{e,r,\omega} + i\,\varepsilon_{e,c,\omega}$) for electronic transition across a bandgap is

$$\epsilon_{e,c,\omega} = \frac{4\pi^2}{\omega^2 V} \sum_{i\in VB,\, j\in CB} \sum_{\kappa} w_\kappa \,|\, p_{ij} \,|^2\, \delta(E_{\kappa,j} - E_{\kappa,i} - \hbar\omega),$$

where V is the unit-cell volume, VB and CB denote the valence and conduction bands, w_κ is the weight associated with a κ-point, and p_{ij} is the transition momentum matrix element. The real part is $\varepsilon_{e,r,\omega}$ is obtained from $\varepsilon_{e,c,\omega}$ using the Kramers-Kronig relation

$$\epsilon_{e,r,\omega} = 1 + \frac{4}{\pi}\mathbb{P}\int_0^\infty d\omega' \frac{\omega' \epsilon_{e,c,\omega'}}{\omega'^2 - \omega^2}.$$

Here, $\mathbb{P}$ denotes the principal value of the integral.

In another example, for the far IR regions where the optical phonons are involved, the dielectric function ($\varepsilon_{e,\omega}$) are calculated as

$$\frac{\epsilon_{e,\omega}}{\epsilon_{e,\infty}} = 1 + \sum_j \frac{\omega_{LO,j}^2 - \omega_{TO,j}^2}{\omega_{TO,j}^2 - \omega^2 - i\gamma\omega},$$

where LO and TO denote the longitudinal and transverse optical phonon modes, j is all the IR-active modes, and γ is the temperature-dependent damping term in the oscillator model. $\varepsilon_{e,\infty}$ is high frequency dielectric permittivity, which can be calculated DFT calculation when ions are treated as external potential.

From these dielectric function ($\varepsilon_{e,\omega}$) calculations (e.g., Abinit, VASP, etc.), the complex refractive index $m_\omega (= n_\omega + i\,\kappa_\omega$, n_ω: refraction index and κ_ω: extinction index) is found, i.e., $m_\omega^{\,2} = \varepsilon_{e,\omega} = \varepsilon_{e,r,\omega} + i\,\varepsilon_{e,c,\omega}$). The surface reflectance R of an ideal surface with normal incident from vacuum or air is given as $R = [(n_\omega - 1)^2 + \kappa_\omega^{\,2}]/[(n_\omega + 1)^2 + \kappa_\omega^{\,2}]$. The spectral absorption coefficient is then found from $\sigma_{ph,\omega} = 2\omega\,\kappa_\omega / u_{ph}$. The spectral absorption coefficient for various electric entities are listed in the below table.

Mechanism	Relation ($\sigma_{ph,\omega}$)
Electronic absorption transition (atom, ion or molecule)	$\frac{n_{e,A}\pi^2 u_{ph}^2 \dot{\gamma}_{ph,e,sp}}{\omega_{e,g}^2 \int_\omega d\omega} = \frac{n_{e,A}\pi\omega_{e,g} \mid \mu_e \mid^2}{3\epsilon_o \hbar u_{ph} \int_\omega d\omega}$, [$n_{e,A}$: number density of ground state, $\omega_{e,g}$: transition angular frequency, $\dot{\gamma}_{ph,e,sp}$: spontaneous emission rate (s^{-1}), $\boldsymbol{\mu}_e$: transition dipole moment, $\int_\omega d\omega$: bandwidth]
Free carrier absorption (metal)	$\frac{2\omega\kappa_\omega}{u_{ph}} = (\frac{2n_{e,c}e_c^2\langle\langle\tau_e\rangle\rangle\omega}{\epsilon_o\epsilon_e m_e u_{ph}^2})^{1/2}$ ($n_{e,c}$: number density of conduction electrons, $\langle\langle\tau_e\rangle\rangle$: average momentum electron relaxation time, ε_o: free space electrical permittivity)
Direct-band absorption (semiconductor)	$\frac{e_c^2 \mid\langle\varphi_v \mid e_c \mathbf{x} \mid \varphi_c\rangle\mid^2 D_{ph-e}[f_e^o(E_{e,v}) - f_e^o(E_{e,c})]}{\epsilon_o^2 m_{e,e}^2 u_{ph} n_\omega \omega}$ (n_ω: index of refraction, $D_{ph\text{-}e}$: joint density of states)
Indirect-band absorption (semiconductor)	with phonon absorption: $\frac{a_{ph-e-p,a}(\hbar\omega - \Delta E_{e,g} + \hbar\omega_p)^2}{exp(\hbar\omega_p / k_B T) - 1}$ ($a_{ph\text{-}e\text{-}p,a}$ phonon absorption coupling coefficient, $\Delta E_{e,g}$: bandgap, ω_p: phonon energy) with phonon emission: $\frac{a_{ph-e-p,e}(\hbar\omega - \Delta E_{e,g} - \hbar\omega_p)^2}{1 - exp(-\hbar\omega_p / k_B T)}$ ($a_{ph\text{-}e\text{-}p,e}$ phonon emission coupling coefficient)

Heat Transfer Fluid

Coolant

Coolant, or antifreeze, is generally a half-and-half mixture of a form of glycol and water. The glycol represents the antifreeze element of the mix, guaranteeing that the fluid doesn't turn into ice under harsh winter conditions. Glycol also prevents the coolant from reaching the boiling point in extreme heat. It keeps engine temperatures stable under all climate extremes and driving conditions.

Pure water actually transfers heat better than coolant (that's why you see straight water used in the radiators of some types of race cars). However, coolant/antifreeze includes additional additives that prevent rust and corrosion in the radiator, engine and vehicle's heater.

Different Types of Antifreeze

Antifreeze is generally comprised of one of two types of glycol:

- Ethylene glycol antifreeze: Until recently, the most common glycol in antifreeze was ethylene glycol, a toxic material that can cause birth defects, reproductive damage or even death if ingested and requires very specific handling. Ethylene glycol antifreeze has a sweet odor and flavor, which makes it dangerously appealing to animals and/or small children.
- Propylene glycol antifreeze: An alternative antifreeze base is propylene glycol. There is very little difference in the performance of either substance – the advantage is the toxicity level. Propylene glycol antifreeze is significantly less toxic than ethylene glycol. This doesn't mean children or pets can ingest it without harm, but, like alcohol, propylene glycol is not toxic at low levels.

How to Dispose of Coolant

Any antifreeze, whether ethylene or propylene glycol based, picks up heavy-metal contamination during use. For this reason, special care must be taken to dispose of used antifreeze. It's safer to have a repair facility flush your cooling system, since they are required by law to dispose of the material safely.

Most communities have procedures for disposing of hazardous waste; if you do your own repairs and maintenance, take advantage of these procedures. Don't pour coolant down your sink or into storm drains.

Coolant Maintenance

Like any other engine fluid, the coolant needs to be checked on a regular basis. You're checking for two things: quantity and condition.

Most vehicles have a coolant recovery tank or overflow reservoir, which makes checking the fluid level a lot easier and safer. The configuration of the radiator and tank/reservoir lets hot coolant expand into the tank as the engine temperature rises. When the engine cools down, a slight vacuum forms in the radiator and the fluid is drawn out of the tank/reservoir and back into the radiator. As long as the radiator cap remains sealed, the coolant can expand and contract without losing a drop.

How to Check Coolant Level

You can check your coolant level simply by looking at the overflow tank. There are two level indicators on the side of the tank: one indicates the safe level when the engine is hot; the other, when it is cold.

If your coolant level is slightly low, it's safe to add a few ounces of plain water to bring the level back up to the appropriate mark. If you have to add more than a quart of liquid to the cooling system, use a glycol/water antifreeze mixture.

Some vehicles' recovery tanks are pressurized when the engine is hot, making the caps as dangerous to remove as radiator caps. Pressurized recovery tanks are clearly marked with warning decals and their cap is a system pressure cap, rather than a simple plug or twist-off cap.

Gases

Air is a common form of a coolant. Air cooling uses either convective airflow (passive cooling), or a forced circulation using fans.

Hydrogen is used as a high-performance gaseous coolant. Its thermal conductivity is higher than all other gases, it has high specific heat capacity, low density and therefore low viscosity, which is an advantage for rotary machines susceptible to windage losses. Hydrogen-cooled turbogenerators are currently the most common electrical generators in large power plants.

Inert gases are used as coolants in gas-cooled nuclear reactors. Helium has a low tendency to absorb neutrons and become radioactive. Carbon dioxide is used in Magnox and AGR reactors.

Sulfur hexafluoride is used for cooling and insulating of some high-voltage power systems (circuit breakers, switches, some transformers, etc.).

Steam can be used where high specific heat capacity is required in gaseous form and the corrosive properties of hot water are accounted for.

Liquids

Device to measure the temperature to which the coolant protects the car from freezing

The most common coolant is water. Its high heat capacity and low cost makes it a suitable heat-transfer medium. It is usually used with additives, like corrosion inhibitors and antifreeze. Antifreeze, a solution of a suitable organic chemical (most often ethylene glycol, diethylene glycol, or propylene glycol) in water, is used when the water-based coolant has to withstand temperatures below 0 °C, or when its boiling point has to be raised. Betaine is a similar coolant, with the exception that it is made from pure plant juice, and is therefore not toxic or difficult to dispose of ecologically.

Very pure deionized water, due to its relatively low electrical conductivity, is used to cool some electrical equipment, often high-power transmitters and high-power vacuum tubes.

Heavy water is a neutron moderator used in some nuclear reactors; it also has a secondary function as their coolant. Light water reactors, both boiling water and pressurised water reactors the most common type, use ordinary (light) water.

Polyalkylene glycol (PAG) is used as high temperature, thermally stable heat transfer fluids exhibiting strong resistance to oxidation. Modern PAGs can also be non-toxic and non-hazardous.

Cutting fluid is a coolant that also serves as a lubricant for metal-shaping machine tools.

Oils are often used for applications where water is unsuitable. With higher boiling points than water, oils can be raised to considerably higher temperatures (above 100 degrees Celsius) without introducing high pressures within the container or loop system in question.

- Mineral oils serve as both coolants and lubricants in many mechanical gears. Castor oil is also used. Due to their high boiling points, mineral oils are used in portable electric radiator-style space heaters in residential applications, and in closed-loop systems for industrial process heating and cooling. Mineral oil is often used in submerged PC systems as it is non-conductive and therefore won't short circuit or damage any parts.
- Silicone oils and fluorocarbon oils (like fluorinert) are favored for their wide range of operating temperatures. However their high cost limits their applications.
- Transformer oil is used for cooling and additional electric insulation of high-power electric transformers.

Waterless coolant is used as an alternative to conventional water and ethylene glycol coolants. With higher boiling points than water (around 370F), the cooling technology resists boil over. The liquid also prevents corrosion.

Fuels are frequently used as coolants for engines. A cold fuel flows over some parts of the engine, absorbing its waste heat and being preheated before combustion. Kerosene and other jet fuels frequently serve in this role in aviation engines.

Freons were frequently used for immersive cooling of e.g. electronics.

Refrigerants are coolants used for reaching low temperatures by undergoing phase change between liquid and gas. Halomethanes were frequently used, most often R-12 and R-22, but due to environmental concerns are being phased out, often with liquified propane or other haloalkanes like R-134a. Anhydrous ammonia is frequently used in large commercial systems, and sulfur dioxide was used in early mechanical refrigerators. Carbon dioxide (R-744) is used as a working fluid in climate control systems for cars, residential air conditioning, commercial refrigeration, and vending machines.

Heat pipes are a special application of refrigerants.

Molten Metals and Salts

Liquid fusible alloys can be used as coolants in applications where high temperature stability is required, e.g. some fast breeder nuclear reactors. Sodium (in sodium cooled fast reactors) or sodium-potassium alloy NaK are frequently used; in special cases lithium can be employed. Another liquid metal used as a coolant is lead, in e.g. lead cooled fast reactors, or a lead-bismuth alloy. Some early fast neutron reactors used mercury.

For certain applications the stems of automotive poppet valves may be hollow and filled with sodium to improve heat transport and transfer.

For very high temperature applications, e.g. molten salt reactors or very high temperature reactors, molten salts can be used as coolants. One of the possible combinations is the mix of sodium fluoride and sodium tetrafluoroborate (NaF-$NaBF_4$). Other choices are FLiBe and FLiNaK.

Liquid Gases

Liquified gases are used as coolants for cryogenic applications, including cryo-electron microscopy, overclocking of computer processors, applications using superconductors, or extremely sensitive sensors and very low-noise amplifiers.

Carbon Dioxide (chemical formula is CO_2) - is used as a coolant replacement for cutting fluids. CO_2 can provide controlled cooling at the cutting interface such that the cutting tool and the workpiece are held at ambient temperatures. The use of CO_2 greatly extends tool life, and on most materials allows the operation to run faster. This is considered a very environmentally friendly method, especially when compared to the use of petroleum oils as lubricants; parts remain clean and dry which often can eliminate secondary cleaning operations.

Liquid nitrogen, which boils at about -196 °C (77K), is the most common and least expensive coolant in use. Liquid air is used to a lesser extent, due to its liquid oxygen content which makes it prone to cause fire or explosions when in contact with combustible materials.

Lower temperatures can be reached using liquified neon which boils at about -246 °C. The lowest temperatures, used for the most powerful superconducting magnets, are reached using liquid helium.

Liquid hydrogen at -250 to -265 °C can also be used as a coolant. Liquid hydrogen is also used both as a fuel and as a coolant to cool nozzles and combustion chambers of rocket engines.

Nanofluids

A new class of coolants are nanofluids which consist of a carrier liquid, such as water, dispersed with tiny nano-scale particles known as nanoparticles. Purpose-designed nanoparticles of e.g. CuO, alumina, titanium dioxide, carbon nanotubes, silica, or metals (e.g. copper, or silver nanorods) dispersed into the carrier liquid enhance the heat transfer capabilities of the resulting coolant compared to the carrier liquid alone. The enhancement can be theoretically as high as 350%. The experiments however did not prove so high thermal conductivity improvements, but found significant increase of the critical heat flux of the coolants.

Some significant improvements are achievable; e.g. silver nanorods of 55±12 nm diameter and 12.8 μm average length at 0.5 vol.% increased the thermal conductivity of water by 68%, and 0.5 vol.% of silver nanorods increased thermal conductivity of ethylene glycol based coolant by 98%. Alumina nanoparticles at 0.1% can increase the critical heat flux of water by as much as 70%; the particles form rough porous surface on the cooled object, which encourages formation of new bubbles, and their hydrophilic nature then helps pushing them away, hindering the formation of the steam layer. Nanofluid with the concentration more than 5% acts like non-Newtonian fluids.

Solids

In some applications, solid materials are used as coolants. The materials require high energy to vaporize; this energy is then carried away by the vaporized gases. This approach is common in spaceflight, for ablative atmospheric reentry shields and for cooling of rocket engine nozzles. The same approach is also used for fire protection of structures, where ablative coating is applied.

Dry ice and water ice can be also used as coolants, when in direct contact with the structure being cooled.

Sublimation of water ice was used for cooling the space suit for Project Apollo.

Transformer Oil

Transformer oil or insulating oil is an oil that is stable at high temperatures and has excellent electrical insulating properties. It is used in oil-filled transformers, some types of high-voltage capacitors, fluorescent lamp ballasts, and some types of high-voltage switches and circuit breakers. Its functions are to insulate, suppress corona discharge and arcing, and to serve as a coolant.

Transformer oil is most often based on mineral oil, but alternative formulations with better engineering or environmental properties are growing in popularity.

Function and Properties

Oil transformer with air convection cooled heat exchangers in the front and at the side

Transformer oil's primary functions are to insulate and cool a transformer. It must therefore have high dielectric strength, thermal conductivity, and chemical stability, and must keep these properties when held at high temperatures for extended periods. Typical specifications are: flash point 140 °C or greater, pour point −30 °C or lower, dielectric breakdown voltage 28 kV (RMS) or greater.

To improve cooling of large power transformers, the oil-filled tank may have external radiators through which the oil circulates by natural convection. Power transformers with capacities of thousands of kVA may also have cooling fans, oil pumps, and even oil-to-water heat exchangers.

Power transformers undergo prolonged drying processes, using electrical self-heating, the application of a vacuum, or both to ensure that the transformer is completely free of water vapor before the insulating oil is introduced. This helps prevent corona formation and subsequent electrical breakdown under load.

Oil filled transformers with a conservator (oil reservoir) may have a gas detector relay (Buchholz relay). These safety devices detect the buildup of gas inside the transformer due to corona discharge, overheating, or an internal electric arc. On a slow accumulation of gas, or rapid pressure rise, these devices can trip a protective circuit breaker to remove power from the transformer. Transformers without conservators are usually equipped with sudden pressure relays, which perform a similar function as the Buchholz relay.

Mineral Oil Alternatives

Mineral oil is generally effective as a transformer oil, but it has some serious disadvantages, of which the worst is its high flammability. If a transformer leaks mineral oil, it can easily start a fire. Fire codes often require that transformers inside buildings use a less flammable liquid, or the use of dry-type transformers with no liquid at all. Mineral oil is also an environmental contaminant, and its insulating properties are rapidly degraded by even small amounts of water.

Pentaerythritol tetra fatty acid natural and synthetic esters have emerged as an increasingly common mineral oil alternative, especially in high-fire-risk applications such as indoors or offshore, due to their low volatility and high fire point, which can be over 300 °C. They also have a lower pour point, greater moisture tolerance, and improved function at high temperatures, and they are non-toxic and readily biodegradable. Silicone or fluorocarbon-based oils, which are even less flammable, are also used, but they are more expensive than esters, and less biodegradable.

A 380kV transformer with vegetable oil

Researchers are experimenting with vegetable-based formulations, using coconut oil for instance. As yet these are unsuitable for use in cold climates or for voltages over 230 kV. Researchers are also investigating nanofluids for transformer use; these would be used as additives to improve the stability and thermal and electrical properties of the oil.

Polychlorinated Biphenyls (PCBs)

Polychlorinated biphenyls (PCBs) were formerly used as transformer oil, since they have high dielectric strength and are not flammable. Unfortunately, they are also toxic, bioaccumulative, not at all biodegradable, and difficult to dispose of safely. When burned, they form even more toxic products, such as chlorinated dioxins and chlorinated dibenzofurans. Beginning in the 1970s, production and new uses of PCBs were banned in many countries, due to concerns about the accumulation of PCBs and toxicity of their byproducts. For instance, in the USA, production of PCBs was banned in 1979 under the Toxic Substances Control Act. In many countries significant programs are in place to reclaim and safely destroy PCB contaminated equipment. One method that can be used to reclaim PCB contaminated transformer oil is the application of a PCB removal system, also called a PCB dechlorination system. PCB removal systems use an alkali dispersion to strip the chlorine atoms from the other molecules in a chemical reaction. This forms PCB-free transformer oil and a PCB-free sludge. The two can then be separated via a centrifuge. The sludge can be disposed as regular non-PCB industrial waste. The treated transformer oil is fully restored, meeting the required standards, without any detectable PCB content. It can, thus, be used as the insulating fluid in transformers again.

PCBs and mineral oil are miscible in all proportions, and sometimes the same equipment (drums, pumps, hoses, and so on) was used for either type of liquid, so PCB contamination of transformer oil continues to be a concern. For instance, under present regulations, concentrations of PCBs exceeding 5 parts per million can cause an oil to be classified as hazardous waste in California.

Testing and Oil Quality

Transformer oils are subject to electrical and mechanical stresses while a transformer is in operation. In addition there is contamination caused by chemical interactions with windings and other solid insulation, catalyzed by high operating temperature. The original chemical properties of transformer oil change gradually, rendering it ineffective for its intended purpose after many years. Oil in large transformers and electrical apparatus is periodically tested for its electrical and chemical properties, to make sure it is suitable for further use. Sometimes oil condition can be improved by filtration and treatment. Tests can be divided into:

1. Dissolved gas analysis
2. Furan analysis
3. PCB analysis
4. General electrical & physical tests:
    - Color & Appearance
    - Breakdown Voltage
    - Water Content
    - Acidity (Neutralization Value)
    - Dielectric Dissipation Factor
    - Resistivity
    - Sediments & Sludge

- Flash Point
- Pour Point
- Density
- Kinematic Viscosity

The details of conducting these tests are available in standards released by IEC, ASTM, IS, BS, and testing can be done by any of the methods. The Furan and DGA tests are specifically not for determining the quality of transformer oil, but for determining any abnormalities in the internal windings of the transformer or the paper insulation of the transformer, which cannot be otherwise detected without a complete overhaul of the transformer. Suggested intervals for these test are:

- General and physical tests - bi-yearly
- Dissolved gas analysis - yearly
- Furan testing - once every 2 years, subject to the transformer being in operation for min 5 years.

On-site Testing

Some transformer oil tests can be carried out in the field, using portable test apparatus. Other tests, such as dissolved gas, normally require a sample to be sent to a laboratory. Electronic on-line dissolved gas detectors can be connected to important or distressed transformers to continually monitor gas generation trends.

To determine the insulating property of the dielectric oil, an oil sample is taken from the device under test, and its breakdown voltage is measured on-site according to the following test sequence:

- In the vessel, two standard-compliant test electrodes with a typical clearance of 2.5 mm are surrounded by the insulating oil.
- During the test, a test voltage is applied to the electrodes. The test voltage is continuously increased up to the breakdown voltage with a constant slew rate of e.g. 2 kV/s.
- Breakdown occurs in an electric arc, leading to a collapse of the test voltage.
- Immediately after ignition of the arc, the test voltage is switched off automatically.
- Ultra fast switch off is crucial, as the energy that is brought into the oil and is burning it during the breakdown, must be limited to keep the additional pollution by carbonisation as low as possible.

- The root mean square value of the test voltage is measured at the very instant of the breakdown and is reported as the breakdown voltage.
- After the test is completed, the insulating oil is stirred automatically and the test sequence is performed repeatedly.
- The resulting breakdown voltage is calculated as mean value of the individual measurements.

Nanofluid

Nanofluids are a new class of fluids engineered by dispersing nanometer-sized materials (nanoparticles, nanofibers, nanotubes, nanowires, nanorods, nanosheet, or droplets) in base fluids. In other words, nanofluids are nanoscale colloidal suspensions containing condensed nanomaterials. They are two-phase systems with one phase (solid phase) in another (liquid phase). Nanofluids have been found to possess enhanced thermophysical properties such as thermal conductivity, thermal diffusivity, viscosity, and convective heat transfer coefficients compared to those of base fluids like oil or water. It has demonstrated great potential applications in many fields.

For a two-phase system, there are some important issues we have to face. One of the most important issues is the stability of nanofluids, and it remains a big challenge to achieve desired stability of nanofluids.

Preparation Methods for Nanofluids

Two-Step Method

Two-step method is the most widely used method for preparing nanofluids. Nanoparticles, nanofibers, nanotubes, or other nanomaterials used in this method are first produced as dry powders by chemical or physical methods. Then, the nanosized powder will be dispersed into a fluid in the second processing step with the help of intensive magnetic force agitation, ultrasonic agitation, high-shear mixing, homogenizing, and ball milling. Two-step method is the most economic method to produce nanofluids in large scale, because nanopowder synthesis techniques have already been scaled up to industrial production levels. Due to the high surface area and surface activity, nanoparticles have the tendency to aggregate. The important technique to enhance the stability of nanoparticles in fluids is the use of surfactants. However, the functionality of the surfactants under high temperature is also a big concern, especially for high-temperature applications.

Due to the difficulty in preparing stable nanofluids by two-step method, several advanced techniques are developed to produce nanofluids, including one-step method. In the following part, we will introduce one-step method in detail.

One-Step Method

To reduce the agglomeration of nanoparticles, Eastman et al. developed a one-step physical vapor condensation method to prepare Cu/ethylene glycol nanofluids. The one-step process consists of simultaneously making and dispersing the particles in the fluid. In this method, the processes of drying, storage, transportation, and dispersion of nanoparticles are avoided, so the agglomeration of nanoparticles is minimized, and the stability of fluids is increased . The one-step processes can prepare uniformly dispersed nanoparticles, and the particles can be stably suspended in the base fluid. The vacuum-SANSS (submerged arc nanoparticle synthesis system) is another efficient method to prepare nanofluids using different dielectric liquids. The different morphologies are mainly influenced and determined by various thermal conductivity properties of the dielectric liquids. The nanoparticles prepared exhibit needle-like, polygonal, square, and circular morphological shapes. The method avoids the undesired particle aggregation fairly well.

One-step physical method cannot synthesize nanofluids in large scale, and the cost is also high, so the one-step chemical method is developing rapidly. Zhu et al. presented a novel one-step chemical method for preparing copper nanofluids by reducing $CuSO \cdot 5H_2O$ with $NaH_2PO_2 \cdot H_2O$ in ethylene glycol under microwave irradiation. Well-dispersed and stably suspended copper nanofluids were obtained. Mineral oil-based nanofluids containing silver nanoparticles with a narrow-size distribution were also prepared by this method . The particles could be stabilized by Korantin, which coordinated to the silver particle surfaces via two oxygen atoms forming a dense layer around the particles. The silver nanoparticle suspensions were stable for about 1 month. Stable ethanol-based nanofluids containing silver nanoparticles could be prepared by microwave-assisted one-step method. In the method, polyvinylpyrrolidone (PVP) was employed as the stabilizer of colloidal silver and reducing agent for silver in solution. The cationic surfactant octadecylamine (ODA) is also an efficient phase-transfer agent to synthesize silver colloids. The phase transfer of the silver nanoparticles arises due to coupling of the silver nanoparticles with the ODA molecules present in organic phase via either coordination bond formation or weak covalent interaction. Phase transfer method has been developed for preparing homogeneous and stable graphene oxide colloids. Graphene oxide nanosheets (GONs) were successfully transferred from water to n-octane after modification by oleylamine, and the schematic illustration of the phase transfer process is shown in figure.

However, there are some disadvantages for one-step method. The most important one is that the residual reactants are left in the nanofluids due to incomplete reaction or stabilization. It is difficult to elucidate the nanoparticle effect without eliminating this impurity effect.

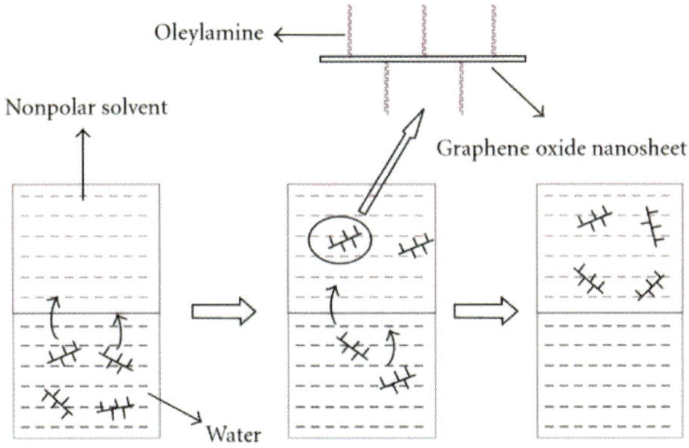

Schematic illustration of the phase transfer process

Other Novel Methods

Wei et al. developed a continuous-flow microfluidic microreactor to synthesize copper nanofluids. By this method, copper nanofluids can be continuously synthesized, and their microstructure and properties can be varied by adjusting parameters such as reactant concentration, flow rate, and additive. CuO nanofluids with high solid volume fraction (up to 10 vol%) can be synthesized through a novel precursor transformation method with the help of ultrasonic and microwave irradiation . The precursor $Cu(OH)_2$ is completely transformed to CuO nanoparticle in water under microwave irradiation. The ammonium citrate prevents the growth and aggregation of nanoparticles, resulting in a stable CuO aqueous nanofluid with higher thermal conductivity than those prepared by other dispersing methods. Phase-transfer method is also a facile way to obtain monodisperse noble metal colloids . In a water-cyclohexane two-phase system, aqueous formaldehyde is transferred to cyclohexane phase via reaction with dodecylamine to form reductive intermediates in cyclohexane. The intermediates are capable of reducing silver or gold ions in aqueous solution to form dodecylamine-protected silver and gold nanoparticles in cyclohexane solution at room temperature. Feng et al. used the aqueous organic phase-transfer method for preparing gold, silver, and platinum nanoparticles on the basis of the decrease of the PVP's solubility in water with the temperature increase . Phase-transfer method is also applied for preparing stable kerosene-based Fe_3O_4 nanofluids. Oleic acid is successfully grafted onto the surface of Fe_3O_4 nanoparticles by chemisorbed mode, which lets Fe_3O_4 nanoparticles have good compatibility with kerosene . The Fe_3O_4 nanofluids prepared by phase-transfer method do not show the previously reported "time dependence of the thermal conductivity characteristic". The preparation of nanofluids with controllable microstructure is one of the key issues. It is well known that the properties of nanofluids strongly depend on the structure and shape of nanomaterials. The recent research shows that nanofluids synthesized by chemical solution method have both higher conductivity enhancement and better stability than those produced by the other methods . This method is distinguished from the others by its controllability. The nanofluid microstructure can be varied and manipulated by adjusting synthesis parameters such as temperature, acidity, ultrasonic and microwave irradiation, types and concentrations of reactants and additives, and the order in which the additives are added to the solution.

Synthesis

Nanofluids are produced by several techniques:

1. Direct Evaporation (1 step)
2. Gas condensation/dispersion (2 step)
3. Chemical vapour condensation (1 step)
4. Chemical precipitation (1 step)

Several liquids including water, ethylene glycol, and oils have been used as base fluids. Although stabilization can be a challenge, on-going research indicates that it is possible. Nano-materials used so far in nanofluid synthesis include metallic particles, oxide particles, carbon nanotubes, graphene nano-flakes and ceramic particles.

Smart Cooling Nanofluids

Realizing the modest thermal conductivity enhancement in conventional nanofluids, a team of researchers at Indira Gandhi Centre for Atomic Research Centre, Kalpakkam developed a new class of magnetically polarizable nanofluids where the thermal conductivity enhancement up to 300% of basefluids is demonstrated. Fatty-acid-capped magnetite nanoparticles of different sizes (3-10 nm) have been synthesized for this purpose. It has been shown that both the thermal and rheological properties of such magnetic nanofluids are tunable by varying the magnetic field strength and orientation with respect to the direction of heat flow. Such response stimuli fluids are reversibly switchable and have applications in miniature devices such as micro- and nano-electromechanical systems. In 2013, Azizian et al. considered the effect of an external magnetic field on the convective heat transfer coefficient of water-based magnetite nanofluid experimentally under laminar flow regime. Up to 300% enhancement obtained at Re=745 and magnetic field gradient of 32.5 mT/mm. The effect of the magnetic field on the pressure drop was not as significant.

Nanoparticle Migration

In nanofluids, it is recognized that nanoparticles do not follow the fluid streamlines passively. In fact, there are some reasons that induce a slip velocity between the nanoparticles and the base fluid. Movements of nanoparticles has significant impact on rheological and thermophysical properties of the nanofluids. Therefore, investigating the nanoparticles motion is critical for evaluating the performance of nanoparticles inclusion to the base fluid as a heat transfer medium. Since the nanoparticles are very small (100 nm), Brownian and thermophoretic diffusivities are the main slip mechanisms in nanofluids, as Buongiorno declared. Brownian diffusion is due to random drifting of suspended nanoparticles in the base fluid which originates from continuous collisions among the nanoparticles and liquid molecules. Thermophoresis induces nanoparticle migration from warmer to colder region (in opposite direction of the temperature gradient), making a non-uniform nanoparticle volume fraction distribution.

In fact, theoretical models estimated that nanoparticles are non-homogeneously distributed. The level of non-uniformity is completely depend on thermal boundary conditions, the nanoparticle size, shape, and material. Rigorous readers encouraged to find more interesting results in open literature.

Response Stimuli Nanofluids for Sensing Applications

Researchers have invented a nanofluid-based ultrasensitive optical sensor that changes its colour on exposure to extremely low concentrations of toxic cations. The sensor is useful in detecting minute traces of cations in industrial and environmental samples. Existing techniques for monitoring cations levels in industrial and environmental samples are expensive, complex and time-consuming. The sensor is designed with a magnetic nanofluid that consists of nano-droplets with magnetic grains suspended in water. At a fixed magnetic field, a light source illuminates the nanofluid where the colour of the nanofluid changes depending on the cation concentration. This color change occurs within a second after exposure to cations, much faster than other existing cation sensing methods.

Such response stimulus nanofluids are also used to detect and image defects in ferromagnetic components. The photonic eye, as it has been called, is based on a magnetically polarizable nano-emulsion that changes colour when it comes into contact with a defective region in a sample. The device might be used to monitor structures such as rail tracks and pipelines.

Magnetically Responsive Photonic Crystals Nanofluids

Magnetic nanoparticle clusters or magnetic nanobeads with the size 80–150 nanometers form ordered structures along the direction of the external magnetic field with a regular interparticle spacing on the order of hundreds of nanometers resulting in strong diffraction of visible light in suspension.

Nanolubricants

Another word used to describe nanoparticle based suspensions is Nanolubricants. They are mainly prepared using oils used for engine and machine lubrication. So far several materials including metals, oxides and allotropes of carbon have been used to formulate nanolubricants. The addition of nanomaterials mainly enhances the thermal conductivity and anti-wear property of base oils. Although MoS2, graphene, Cu based fluids have been studied extensively, the fundamental understanding of underlying mechanisms is still needed.

Molybdenum disulfide (MoS2) and graphene work as third body lubricants, essentially becoming tiny microscopic ball bearings, which reduce the friction between two contacting surfaces. This mechanism is beneficial if a sufficient supply of these particles are present at the contact interface. The beneficial effects are diminished as the rubbing mechanism pushes out the third body lubricants. Changing the lubricant, like-wise, will null the effects of the nanolubricants drained with the oil.

Other nanolubricant approaches, such as Magnesium Silicate Hydroxides (MSH) rely on nanoparticle coatings by synthesizing nanomaterials with adhesive and lubricating functionalities. Research into nanolubricant coatings has been conducted in both the academic and industrial spaces. Nanoborate additives as well as mechanical model descriptions of diamond-like carbon (DLC) coating formations have been developed by Ali Erdemir at Argonne National Labs. Companies such as TriboTEX provide consumer formulations of synthesized MSH nanomaterial coatings for vehicle engines and industrial applications.

Nanofluids in Petroleum Refining Process

Many researches claim that nanoparticles can be used to enhance crude oil recovery. It is evident that development of nanofluids for oil and gas industry has a great practical aspects.

Thermophysical Properties of Nanofluids

There are four thermophysical properties related to fluid, which are namely viscosity, thermal conductivity, specific heat capacity and density.

Density

Pak and Choi Model

$$\rho_{nf} = \rho_{bf}\left(1 + \frac{\varnothing(\rho_p - \rho_{bf})}{\rho_{bf}}\right)$$

Specific Heat Capacity

Xuan and Roetzel model

$$(C_p)_{nf} = \frac{(C_p)_{bf}\left(1 - \frac{\varnothing\left(((C_p)_{bf}\rho_{bf} - (C_p)_p\rho_p\right)}{(C_p)_{bf}\rho_{bf}}\right)}{\left(1 + \frac{\varnothing(\rho_p - \rho_{bf})}{\rho_p}\right)}$$

Application of Nanofluids

Heat Transfer Intensification

Since the origination of the nanofluid concept about a decade ago, the potentials of nanofluids in heat transfer applications have attracted more and more attention. Up to now, there are some review papers which present overviews of various aspects of nanofluids, including preparation and characterization, techniques for the measurements of thermal conductivity, theory and model, thermophysical properties, and convective heat transfer. Our group studied the thermal conductivities of ethylene glycol- (EG-) based nanofluids containing oxides including MgO, TiO_2, ZnO, Al_2O_3, and SiO_2 nanoparticles, and the results (table below) demonstrated that MgO-EG nanofluid was found to have superior features with the highest thermal conductivity and lowest viscosity. In this part, we will summarize the applications of nanofluids in heat transfer enhancement.

Table: Properties of oxides and their nanofluids.

	Thermal conductivity* W/(m.k)	Density (g/cm³)	Crystalline	Viscosity (Cp) with 5.0 vol.% 30° C	Thermal conductivity enchancement of nanofluids (%) with 5.0 vol. %
MgO	48.4	2.9	Cubic	17.4	40.6
TiO_2	8.4	4.1	Anatase	31.2	27.2
ZnO	13.0	5.6	Wurtzite	129.2	26.8
Al_2O_3	36.0	3.6	Y	28.2	28.2
SiO_3	10.4	2.6	noncrystalline	31.5	25.3

Electronic Applications

Due to higher density of chips, design of electronic components with more compact makes heat dissipation more difficult. Advanced electronic devices face thermal management challenges from the high level of heat generation and the reduction of available surface area for heat removal. So, the reliable thermal management system is vital for the smooth operation of the advanced electronic devices. In general, there are two approaches to improve the heat removal for electronic equipment. One is to find an optimum geometry of cooling devices; another is to increase the heat transfer capacity. Nanofluids with higher thermal conductivities are predicated convective heat transfer coefficients compared to those of base fluids. Recent researches illustrated that nanofluids could increase the heat transfer coefficient by increasing the thermal conductivity of a coolant. Jang and Choi designed a new cooler, combined microchannel heat sink with nanofluids . Higher cooling performance was obtained when compared to the device using pure water as working medium. Nanofluids reduced both the thermal resistance and the temperature difference between the heated microchannel wall and the coolant. A combined microchannel heat sink with nanofluids had the potential as the next-generation cooling devices for removing ultrahigh heat flux. Nguyen et al.

designed a closed liquid-circuit to investigate the heat transfer enhancement of a liquid cooling system by replacing the base fluid (distilled water) with a nanofluid composed of distilled water and Al_2O_3 nanoparticles at various concentrations. Measured data have clearly shown that the inclusion of nanoparticles within the distilled water has produced a considerable enhancement in convective heat transfer coefficient of the cooling block. With particle loading 4.5 vol%, the enhancement is up to 23% with respect to that of the base fluid. It has also been observed that an augmentation of particle concentration has produced a clear decrease of the junction temperature between the heated component and the cooling block. Silicon microchannel heat sink performance using nanofluids containing Cu nanoparticles was analyzed . It was found that nanofluids could enhance the performance as compared with that using pure water as the coolant. The enhancement was due to the increase in thermal conductivity of coolant and the nanoparticle thermal dispersion effect. The other advantage was that there was no extra pressure drop, since the nanoparticle was small, and particle volume fraction was low.

The thermal requirements on the personal computer become much stricter with the increase in thermal dissipation of CPU. One of the solutions is the use of heat pipes. Nanofluids, employed as working medium for conventional heat pipe, have shown higher thermal performances, having the potential as a substitute for conventional water in heat pipe. At a same charge volume, there is a significant reduction in thermal resistance of heat pipe with nanofluid containing gold nanoparticles as compared with water . The measured results also show that the thermal resistance of a vertical meshed heat pipe varies with the size of gold nanoparticles. The suspended nanoparticles tend to bombard the vapor bubble during the bubble formation. Therefore, it is expected that the nucleation size of vapor bubble is much smaller for fluid with suspended nanoparticles than that without them. This may be the major reason for reducing the thermal resistance of heat pipe. Chen et al. studied the effect of a nanofluid on flat heat pipe (FHP) thermal performance, using silver nanofluid as the working fluid. The temperature difference and the thermal resistance of the FHP with the silver nanoparticle solution were lower than those with pure water. The plausible reasons for enhancement of the thermal performance of the FHP using the nanofluid can be explained by the critical heat flux enhancement by higher wettability and the reduction of the boiling limit. Nanofluid oscillating heat pipe with ultrahigh-performance was developed by Ma et al. . They combined nanofluids with thermally excited oscillating motion in an oscillating heat pipe, and heat transport capability significantly increased. For example, at the input power of 80.0 W, diamond nanofluid could reduce the temperature difference between the evaporator and the condenser from 40.9 to 24.3°C. This study would accelerate the development of a highly efficient cooling device for ultrahigh-heat-flux electronic systems. The thermal performance investigation of heat pipe indicated that nanofluids containing silver or titanium nanoparticles could be used as an efficient cooling fluid for devices with high energy density. For a silver nanofluid, the temperature difference decreased 0.56–0.65 compared to water at an input power of 30–50

W . For the heat pipe with titanium nanoparticles at a volume concentration of 0.10%, the thermal efficiency is 10.60% higher than that with the based working fluid . These positive results are promoting the continued research and development of nanofluids for such applications.

Transportation

Nanofluids have great potentials to improve automotive and heavy-duty engine cooling rates by increasing the efficiency, lowering the weight and reducing the complexity of thermal management systems. The improved cooling rates for automotive and truck engines can be used to remove more heat from higher horsepower engines with the same size of cooling system. Alternatively, it is beneficial to design more compact cooling system with smaller and lighter radiators. It is, in turn, beneficial the high performance and high fuel economy of car and truck. Ethylene glycol-based nanofluids have attracted much attention in the application as engine coolant [56–58] due to the low-pressure operation compared with a 50/50 mixture of ethylene glycol and water, which is the nearly universally used automotive coolant. The nanofluids has a high boiling point, and it can be used to increase the normal coolant operating temperature and then reject more heat through the existing coolant system . Kole et al. prepared car engine coolant (Al_2O_3 nanofluid) using a standard car engine coolant (HP KOOLGARD) as the base fluid and studied the thermal conductivity and viscosity of the coolant. The prepared nanofluid, containing only 3.5% volume fraction of Al_2O_3 nanoparticles, displayed a fairly higher thermal conductivity than the base fluid, and a maximum enhancement of 10.41% was observed at room temperature. Tzeng et al. applied nanofluids to the cooling of automatic transmissions. The experimental platform was the transmission of a four-wheel drive vehicle. The used nanofluids were prepared by dispersing CuO and Al_2O_3 nanoparticles into engine transmission oil. The results showed that CuO nanofluids produced the lower transmission temperatures both at high and low rotating speeds. From the thermal performance viewpoint, the use of nanofluid in the transmission has a clear advantage.

The researchers of Argonne National Laboratory have assessed the applications of nanofluids for transportation . The use of high-thermal conductive nanofluids in radiators can lead to a reduction in the frontal area of the radiator up to 10%. The fuel saving is up to 5% due to the reduction in aerodynamic drag. It opens the door for new aerodynamic automotive designs that reduce emissions by lowering drag. The application of nanofluids also contributed to a reduction of friction and wear, reducing parasitic losses, operation of components such as pumps and compressors, and subsequently leading to more than 6% fuel savings. In fact, nanofluids not only enhance the efficiency and economic performance of car engine, but also will greatly influence the structure design of automotives. For example, the engine radiator cooled by a nanofluid will be smaller and lighter. It can be placed elsewhere in the vehicle, allowing for the redesign of a far more aerodynamic chassis. By reducing the size and changing the location of the radiator, a reduction in weight and wind resistance could enable greater fuel efficiency and subsequently lower exhaust emissions. Computer simulations from the US department of energy's office of vehicle technology showed that nanofluid coolants could reduce the size of truck radiators by 5%. This would result in a 2.5% fuel saving at highway speeds.

The practical applications are on the road. In USA, car manufacturers GM and Ford are running their own research programs on nanofluid applications. A €8.3 million FP7 project, named NanoHex (Nanofluid Heat Exchange), began to run. It involved 12 organizations from Europe and Israel ranging from Universities to SMEs and major companies. NanoHex is overcoming the technological challenges faced in development and application of reliable and safe nanofluids for more sophisticated, energy efficient, and environmentally friendly products and services .

Industrial Cooling Applications

The application of nanofluids in industrial cooling will result in great energy savings and emissions reductions. For US industry, the replacement of cooling and heating water with nanofluids has the potential to conserve 1 trillion Btu of energy [41, 64]. For the US electric power industry, using nanofluids in closed loop cooling cycles could save about 10–30 trillion Btu per year (equivalent to the annual energy consumption of about 50,000–150,000 households). The associated emissions reductions would be approximately 5.6 million metric tons of carbon dioxide, 8,600 metric tons of nitrogen oxides, and 21,000 metric tons of sulfur dioxide .

Experiments were performed using a flow-loop apparatus to explore the performance of polyalphaolefin nanofluids containing exfoliated graphite nanoparticle fibers in cooling . It was observed that the specific heat of nanofluids was found to be 50% higher for nanofluids compared with polyalphaolefin, and it increased with temperature. The thermal diffusivity was found to be 4 times higher for nanofluids. The convective heat transfer was enhanced by 10% using nanofluids compared with using polyalphaolefin. Ma et al. proposed the concept of nanoliquid-metal fluid, aiming to establish an engineering route to make the highest conductive coolant with about several dozen times larger thermal conductivity than that of water . The liquid metal with low melting point is expected to be an idealistic base fluid for making superconductive solution, which may lead to the ultimate coolant in a wide variety of heat transfer enhancement area. The thermal conductivity of the liquid-metal fluid can be enhanced through the addition of more conductive nanoparticles.

Heating Buildings and Reducing Pollution

Nanofluids can be applied in the building heating systems. Kulkarni et al. evaluated how they perform heating buildings in cold regions . In cold regions, it is a common practice to use ethylene or propylene glycol mixed with water in different proportions as a heat transfer fluid. So, 60:40 ethylene glcol/water (by weight) was selected as the base fluid. The results showed that using nanofluids in heat exchangers could reduce volumetric and mass flow rates, resulting in an overall pumping power savings. Nanofluids necessitate smaller heating systems, which are capable of delivering the same amount of thermal energy as larger heating systems but are less expensive. This lowers the initial equipment cost excluding nanofluid cost. This will also reduce environmental pollutants, because smaller heating units use less power, and the heat transfer unit has less liquid and material waste to discard at the end of its life cycle.

Nuclear Systems Cooling

The Massachusetts Institute of Technology has established an interdisciplinary center for nanofluid technology for the nuclear energy industry. The researchers are exploring the nuclear applications of nanofluids, specifically the following three : (1) main reactor coolant for pressurized water reactors (PWRs). It could enable significant power uprates in current and future PWRs, thus enhancing their economic performance. Specifically, the use of nanofluids with at least 32% higher critical heat flux (CHF) could enable a 20% power density uprate in current plants without changing the fuel assembly design and without reducing the margin to CHF; (2) coolant for the emergency core cooling systems (ECCSs) of both PWRs and boiling water reactors. The use of a nanofluid in the ECCS accumulators and safety injection can increase the peak-cladding-temperature margins (in the nominal-power core) or maintain them in uprated cores if the nanofluid has a higher post-CHF heat transfer rate; (3) coolant for in-vessel retention of the molten core during severe accidents in high-power-density light water reactors. It can increase the margin to vessel breach by 40% during severe accidents in high-power density systems such as Westinghouse APR1000 and the Korean APR1400. While there exist several significant gaps, including the nanofluid thermal-hydraulic performance at prototypical reactor conditions and the compatibility of the nanofluid chemistry with the reactor materials. Much work should be done to overcome these gaps before any applications can be implemented in a nuclear power plant.

Space and Defense

Due to the restriction of space, energy, and weight in space station and aircraft, there is a strong demand for high efficient cooling system with smaller size. You et al. and Vassalo et al. have reported order of magnitude increases in the critical heat flux in pool boiling with nanofluids compared to the base fluid alone. Further research of nanofluids will lead to the development of next generation of cooling devices that incorporate nanofluids for ultrahigh-heat-flux electronic systems, presenting the possibility of raising chip power in electronic components or simplifying cooling requirements for space applications. A number of military devices and systems require high-heat flux cooling to the level of tens of MW/m^2. At this level, the cooling of military devices and system is vital for the reliable operation. Nanofluids with high critical heat fluxes have the potential to provide the required cooling in such applications as well as in other military systems, including military vehicles, submarines, and high-power laser diodes. Therefore, nanofluids have wide application in space and defense fields, where power density is very high and the components should be smaller and weight less.

Thermal Radiation

Thermal radiation is electromagnetic radiation emitted from a material that is due to the heat of the material, the characteristics of which depend on its temperature. An example of thermal radiation is the infrared radiation emitted by a common household radiator or electric heater. A person near a raging bonfire will feel the radiated heat of the fire, even if the surrounding air is very cold. Thermal radiation is generated when heat from the movement of charges in the material (electrons and protons in common forms of matter) is converted to electromagnetic radiation. Sunshine, or solar radiation, is thermal radiation from the extremely hot gasses of the sun, and this radiation heats the earth. The earth also emits thermal radiation but at a much lower intensity because it is cooler. The balance between heating by incoming solar thermal radiation and cooling by the earth's outgoing thermal radiation is the primary process that determines the earth's overall temperature. As such, radiation is the only form of heat transfer that does not require a material to transmit the heat. Radiative heat is transferred from surface to surface, with little heat absorbed between surfaces. However, the surfaces, once heated, can release the heat via conduction or convection to the surroundings.

Thermal radiation is conducted via electromagnetic waves. As such, this form of heat transfer is not only a function of the temperature difference between the two surfaces but also the frequency range of the emitted and received energy. As an example, sunlight is composed of the visible light spectrum as well as infrared energy and ultraviolet energy.

The characteristics of thermal radiation depend on various properties of the surface it is emanating from, including its temperature, its spectral absorptivity and spectral emissive power, as expressed by Kirchhoff's law. The radiation is not monochromatic, i.e., it does not consist of just a single frequency, but comprises a continuous dispersion of photon energies, its characteristic spectrum. If the radiating body and its surface are in thermodynamic equilibrium and the surface has perfect absorptivity at all wavelengths, it is characterized as a black body. A black body is also a perfect emitter. The radiation of such perfect emitters is called black-body radiation. The ratio of any body's emission relative to that of a black body is the body's emissivity, so that a black body has an emissivity of unity.

Absorptivity, reflectivity, and emissivity of all bodies are dependent on the wavelength of the radiation. Due to reciprocity, absortivity and emissivity for any particular wavelength are equal - a good absorber is necessarily a good emitter, and a poor absorber a poor emitter. The temperature determines the wavelength distribution of the electromagnetic radiation. For example, the white paint in the diagram to the right is highly reflective to visible light (reflectivity about 0.80), and so appears white to the human eye due to reflecting sunlight, which has a peak wavelength of about 0.5 micrometers. However, its emissivity at a temperature of about −5 °C (23 °F), peak wavelength of about 12 micrometers, is 0.95. Thus, to thermal radiation it appears black.

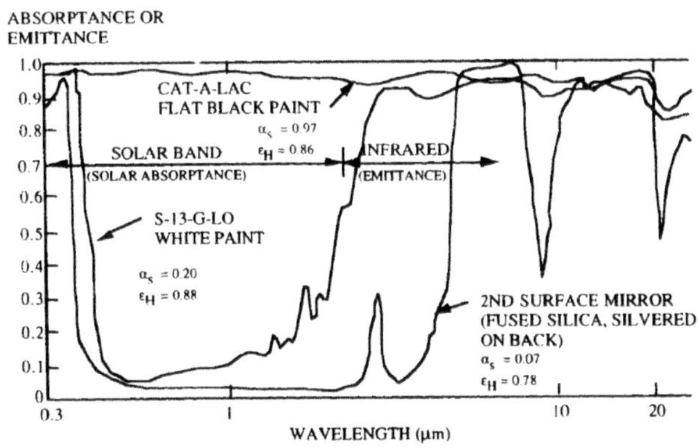

Spectral response of two paints and a mirrored surface, in the visible and the infrared

The distribution of power that a black body emits with varying frequency is described by Planck's law. At any given temperature, there is a frequency f_{max} at which the power emitted is a maximum. Wien's displacement law, and the fact that the frequency is inversely proportional to the wavelength, indicates that the peak frequency f_{max} is proportional to the absolute temperature T of the black body. The photosphere of the sun, at a temperature of approximately 6000 K, emits radiation principally in the (humanly) visible portion of the electromagnetic spectrum. Earth's atmosphere is partly transparent to visible light, and the light reaching the surface is absorbed or reflected. Earth's surface emits the absorbed radiation, approximating the behavior of a black body at 300 K with spectral peak at f_{max}. At these lower frequencies, the atmosphere is largely opaque and radiation from Earth's surface is absorbed or scattered by the atmosphere. Though some radiation escapes into space, most is absorbed and then re-emitted by atmospheric gases. It is this spectral selectivity of the atmosphere that is responsible for the planetary greenhouse effect, contributing to global warming and climate change in general (but also critically contributing to climate stability when the composition and properties of the atmosphere are not changing).

The incandescent light bulb has a spectrum overlapping the black body spectra of the sun and the earth. Some of the photons emitted by a tungsten light bulb filament at 3000 K are in the visible spectrum. Most of the energy is associated with photons of longer wavelengths; these do not help a person see, but still transfer heat to the environment, as can be deduced empirically by observing an incandescent light bulb. Whenever EM radiation is emitted and then absorbed, heat is transferred. This principle is used in microwave ovens, laser cutting, and RF hair removal.

Unlike conductive and convective forms of heat transfer, thermal radiation can be concentrated in a tiny spot by using reflecting mirrors. Concentrating solar power takes advantage of this fact. In many such systems, mirrors are employed to concentrate sunlight into a smaller area. Instead of mirrors, Fresnel lenses can also be used to concentrate heat flux. (In principle, any kind of lens can be used, but only the Fresnel lens design is practical for very large lenses.) Either method can be used to quickly vaporize water into steam using sunlight. For example, the sunlight reflected from mirrors heats the PS10 Solar Power Plant, and during the day it can heat water to 285 °C (558.15 K) or 545 °F.

Surface Effects

Lighter colors and also whites and metallic substances absorb less illuminating light, and thus heat up less; but otherwise color makes small difference as regards heat transfer between an object at everyday temperatures and its surroundings, since the dominant emitted wavelengths are nowhere near the visible spectrum, but rather in the far infrared. Emissivities at those wavelengths have little to do with visual emissivities (visible colors); in the far infra-red, most objects have high emissivities. Thus, except in sunlight, the color of clothing makes little difference as regards warmth; likewise, paint color of houses makes little difference to warmth except when the painted part is sunlit.

The main exception to this is shiny metal surfaces, which have low emissivities both in the visible wavelengths and in the far infrared. Such surfaces can be used to reduce heat transfer in both directions; an example of this is the multi-layer insulation used to insulate spacecraft.

Low-emissivity windows in houses are a more complicated technology, since they must have low emissivity at thermal wavelengths while remaining transparent to visible light.

Nanostructures with spectrally selective thermal emittance properties offer numerous technological applications for energy generation and efficiency, e.g., for cooling photovoltaic cells and buildings. These applications require high emittance in the frequency range corresponding to the atmospheric transparency window in 8 to 13 micron wavelength range. A selective emitter radiating strongly in this range is thus exposed to the clear sky, enabling the use of the outer space as a very low temperature heat sink.

Personalized cooling technology is another example of an application where optical spectral selectivity can be beneficial. Conventional personal cooling is typically achieved through heat conduction and convection. However, the human body is a very efficient emitter of IR radiation, which provides additional cooling mechanism. Most conventional fabrics are opaque to IR radiation and block thermal emission from the body to the environment. Fabrics for personalized cooling applications have been proposed that enable IR transmission to directly pass through clothing, while being opaque at visible wavelengths. Fabrics that are transparent in the infrared can radiate body heat at rates that will significantly reduce the burden on power-hungry air-conditioning systems.

Properties

There are four main properties that characterize thermal radiation (in the limit of the far field):

- Thermal radiation emitted by a body at any temperature consists of a wide range of frequencies. The frequency distribution is given by Planck's law of black-body radiation for an idealized emitter as shown in the diagram at top.
- The dominant frequency (or color) range of the emitted radiation shifts to higher frequencies as the temperature of the emitter increases. For example, a *red hot* object radiates mainly in the long wavelengths (red and orange) of the visible band. If it is heated further, it also begins to emit discernible amounts of green and blue light, and the spread of frequencies in the entire visible range cause it to appear white to the human eye; it is *white hot*. Even at a white-hot temperature of 2000 K, 99% of the energy of the radiation is still in the infrared. This is determined by Wien's displacement law. In the diagram the peak value for each curve moves to the left as the temperature increases.
- The total amount of radiation of all frequencies increases steeply as the temperature rises; it grows as T^4, where T is the absolute temperature of the body. An object at the temperature of a kitchen oven, about twice the room temperature on the absolute temperature scale (600 K vs. 300 K) radiates 16 times as much power per unit area. An object at the temperature of the filament in an incandescent light bulb—roughly 3000 K, or 10 times room temperature—radiates 10,000 times as much energy per unit area. The total radiative intensity of a black body rises as the fourth power of the absolute temperature, as expressed by the Stefan–Boltzmann law. In the plot, the area under each curve grows rapidly as the temperature increases.
- The rate of electromagnetic radiation emitted at a given frequency is proportional to the amount of absorption that it would experience by the source, a property known as reciprocity. Thus, a surface that absorbs more red light thermally radiates more red light. This principle applies to all properties of the wave, including wavelength (color), direction, polarization, and even coherence, so that it is quite possible to have thermal radiation which is polarized, coherent, and directional, though polarized and coherent forms are fairly rare in nature far from sources (in terms of wavelength).

Near-field and Far-field

The general properties of thermal radiation as described by the Planck's law apply if the linear dimension of all parts considered, as well as radii of curvature of all surfaces are large compared with the wavelength of the ray considered' (typically from 8-25 micrometres for the emitter at 300 K). Indeed, thermal radiation as discussed above takes only radiating waves (far-field, or electromagnetic radiation) into account. A more sophisticated framework involving electromagnetic theory must be used for smaller distances from the thermal source or surface (near-field thermal radiation). For example, although far-field thermal radiation at distances from surfaces of more than one wavelength is generally not coherent to any extent, near-field thermal radiation (i.e., radiation at distances of a fraction of various radiation wavelengths) may exhibit a degree of both temporal and spatial coherence.

Planck's law of thermal radiation has been challenged in recent decades by predictions and successful demonstrations of the radiative heat transfer between objects separated by nanoscale gaps that deviate significantly from the law predictions. This deviation is especially strong (up to several orders in magnitude) when the emitter and absorber support surface polariton modes that can couple through the gap separating cold and hot objects. However, to take advantage of the surface-polariton-mediated near-field radiative heat transfer, the two objects need to be separated by ultra-narrow gaps on the order of microns or even nanometers. This limitation significantly complicates practical device designs.

Another way to modify the object thermal emission spectrum is by reducing the dimensionality of the emitter itself. This approach builds upon the concept of confining electrons in quantum wells, wires and dots, and tailors thermal emission by engineering confined photon states in two- and three-dimensional potential traps, including wells, wires, and dots. Such spatial confinement concentrates photon states and enhances thermal emission at select frequencies. To achieve the required level of photon confinement, the dimensions of the radiating objects should be on the order of or below the thermal wavelength predicted by Planck's law. Most importantly, the emission spectrum of thermal wells, wires and dots deviates from Planck's law predictions not only in the near field, but also in the far field, which significantly expands the range of their applications.

Subjective Color to the Eye of a Black Body Thermal Radiator

°C (°F)	Subjective color
480 °C (896 °F)	faint red glow
580 °C (1,076 °F)	dark red
730 °C (1,350 °F)	bright red, slightly orange
930 °C (1,710 °F)	bright orange
1,100 °C (2,010 °F)	pale yellowish orange
1,300 °C (2,370 °F)	yellowish white
> 1,400 °C (2,550 °F)	white (yellowish if seen from a distance through atmosphere)

Selected Radiant Heat Fluxes

The time to a damage from exposure to radiative heat is a function of the rate of delivery of the heat. Radiative heat flux and effects: (1 W/cm^2 = 10 kW/m^2)

kW/m^2	Effect
170	Maximum flux measured in a post-flashover compartment
80	Thermal Protective Performance test for personal protective equipment
52	Fiberboard ignites at 5 seconds
29	Wood ignites, given time
20	Typical beginning of flashover at floor level of a residential room
16	Human skin: sudden pain and second-degree burn blisters after 5 seconds
12.5	Wood produces ignitable volatiles by pyrolysis
10.4	Human skin: Pain after 3 seconds, second-degree burn blisters after 9 seconds
6.4	Human skin: second-degree burn blisters after 18 seconds
4.5	Human skin: second-degree burn blisters after 30 seconds
2.5	Human skin: burns after prolonged exposure, radiant flux exposure typically encountered during firefighting
1.4	Sunlight, sunburns potentially within 30 minutes. Note sunburn is NOT a thermal burn, it is caused by DNA damage due to ultraviolet radiation.

Interchange of Energy

Radiant heat panel for testing precisely quantified energy exposures at National Research Council, near Ottawa, Ontario, Canada

Thermal radiation is one of the three principal mechanisms of heat transfer. It entails the emission of a spectrum of electromagnetic radiation due to an object's temperature.

Other mechanisms are convection and conduction. The interplay of energy exchange by thermal radiation is characterized by the following equation:

$$\alpha + \rho + \tau = 1.$$

Here, α represents the spectral absorption component, ρ spectral reflection component and τ the spectral transmission component. These elements are a function of the wavelength (λ) of the electromagnetic radiation. The spectral absorption is equal to the emissivity ϵ; this relation is known as Kirchhoff's law of thermal radiation. An object is called a black body if, for all frequencies, the following formula applies:

$$\alpha = \epsilon = 1.$$

In a practical situation and room-temperature setting, humans lose considerable energy due to thermal radiation in infra-red in addition to that lost by conduction to air (aided by concurrent convection, or other air movement like drafts). The heat energy lost is partially regained by absorbing heat radiation from walls or other surroundings. (Heat gained by conduction would occur for air temperature higher than body temperature). Otherwise Body temperature is maintined from generated heat through internal metabolism. Human skin has an emissivity of very close to 1.0. Using the formulas below shows a human, having roughly 2 square meter in surface area, and a temperature of about 307 K, continuously radiates approximately 1000 watts. If people are indoors, surrounded by surfaces at 296 K, they receive back about 900 watts from the wall, ceiling, and other surroundings, so the net loss is only about 100 watts. These heat transfer estimates are highly dependent on extrinsic variables, such as wearing clothes, i.e. decreasing total thermal circuit conductivity, therefore reducing total output heat flux. Only truly *gray* systems (relative equivalent emissivity/absorptivity and no directional transmissivity dependence in *all* control volume bodies considered) can achieve reasonable steady-state heat flux estimates through the Stefan-Boltzmann law. Encountering this "ideally calculable" situation is almost impossible (although common engineering procedures surrender the dependency of these unknown variables and "assume" this to be the case). Optimistically, these "gray" approximations will get close to real solutions, as most divergence from Stefan-Boltzmann solutions is very small (especially in most STP lab controlled environments).

If objects appear white (reflective in the visual spectrum), they are not necessarily equally reflective (and thus non-emissive) in the thermal infrared. Most household radiators are painted white but this is sensible given that they are not hot enough to radiate any significant amount of heat, and are not designed as thermal radiators at all - they are actually convectors, and painting them matt black would make little difference to their efficacy. Acrylic and urethane based white paints have 93% blackbody radiation efficiency at room temperature (meaning the term "black body" does not always correspond to the visually perceived color of an object). These materials that do not follow the "black color = high emissivity/absorptivity" caveat will most likely have functional spectral emissivity/absorptivity dependence.

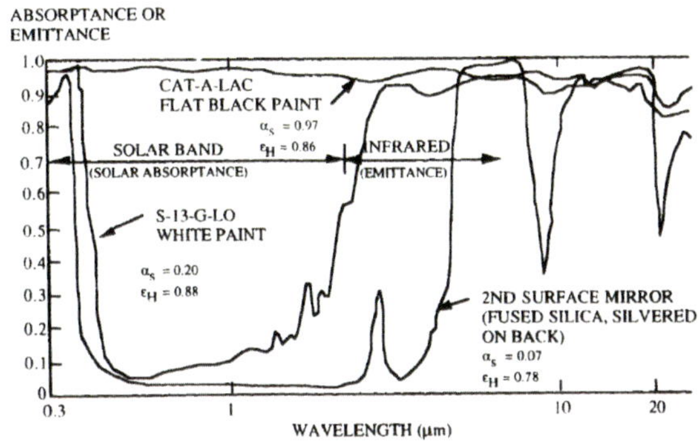

Spectral response of two paints and a mirrored surface, in the visible and the infrared. From NASA.

Calculation of radiative heat transfer between groups of object, including a 'cavity' or 'surroundings' requires solution of a set of simultaneous equations using the radiosity method. In these calculations, the geometrical configuration of the problem is distilled to a set of numbers called view factors, which give the proportion of radiation leaving any given surface that hits another specific surface. These calculations are important in the fields of solar thermal energy, boiler and furnace design and raytraced computer graphics.

A comparison of a thermal image (top) and an ordinary photograph (bottom) shows that a trash bag is transparent but glass (the man's spectacles) is opaque in long-wavelength infrared.

A selective surface can be used when energy is being extracted from the sun. For instance, when a green house is made, most of the roof and walls are made out of glass. Glass is transparent in the visible (approximately $0.4\ \mu m < \lambda < 0.8\ \mu m$) and near-infrared wavelengths, but opaque to mid- to far-wavelength infrared (approximately $\lambda > 3\ \mu m$). Therefore, glass lets in radiation in the visible range, allowing us to be able to see through it, but does not let out radiation that is emitted from objects at or close to room temperature. This traps what we feel as heat. This is known as the greenhouse effect and can be observed by getting into a car that has been sitting in the sun. Selective surfaces can also be used on solar collectors. We can find out how much help a selective surface coating is by looking at the equilibrium temperature of a plate that is being heated through solar radiation. If the plate is receiving a solar irradiation of 1350 W/m² (minimum is 1325 W/m² on July 4 and maximum is 1418 W/m² on January 3) from the sun the temperature of the plate where the radiation leaving is equal to the radiation being received by the plate is 393 K (248 °F). If the plate has a selective surface with an emissivity of 0.9 and a cut off wavelength of 2.0 μm, the equilibrium temperature is approximately 1250 K (1790 °F). The calculations were made neglecting convective heat transfer and neglecting the solar irradiation absorbed in the clouds/atmosphere for simplicity, the theory is still the same for an actual problem.

To reduce the heat transfer from a surface, such as a glass window, a clear reflective film with a low emissivity coating can be placed on the interior of the surface. "Low-emittance (low-E) coatings are microscopically thin, virtually invisible, metal or metallic oxide layers deposited on a window or skylight glazing surface primarily to reduce the U-factor by suppressing radiative heat flow". By adding this coating we are limiting the amount of radiation that leaves the window thus increasing the amount of heat that is retained inside the window.

Since any electromagnetic radiation, including thermal radiation, conveys momentum as well as energy, thermal radiation also induces very small forces on the radiating or absorbing objects. Normally these forces are negligible, but they must be taken into account when considering spacecraft navigation. The Pioneer anomaly, where the motion of the craft slightly deviated from that expected from gravity alone, was eventually tracked down to asymmetric thermal radiation from the spacecraft. Similarly, the orbits of asteroids are perturbed since the asteroid absorbs solar radiation on the side facing the sun, but then re-emits the energy at a different angle as the rotation of the asteroid carries the warm surface out of the sun's view (the YORP effect).

Radiative Power

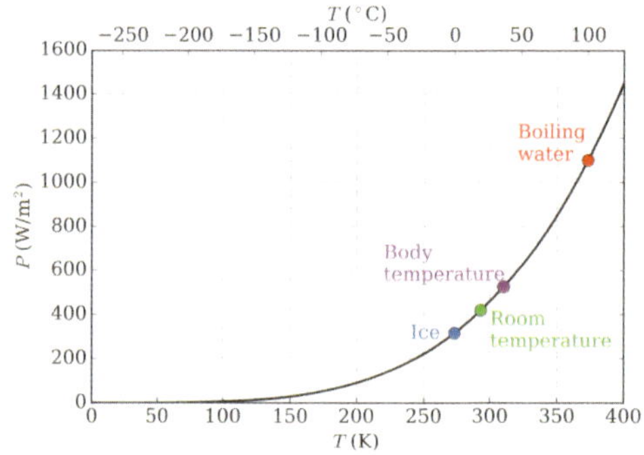

Power emitted by a black body plotted against the temperature according to the Stefan–Boltzmann law.

Thermal radiation power of a black body per unit area of radiating surface per unit of solid angle and per unit frequency ν is given by Planck's law as:

$$u(\nu,T)=\frac{2h\nu^3}{c^2}\cdot\frac{1}{e^{h\nu/k_BT}-1}$$

or instead of per unit frequency, per unit wavelength as

$$u(\lambda,T)=\frac{2hc^2}{\lambda^5}\cdot\frac{1}{e^{hc/k_BT\lambda}-1}$$

This formula mathematically follows from calculation of spectral distribution of energy in quantized electromagnetic field which is in complete thermal equilibrium with the radiating object. The equation is derived as an infinite sum over all possible frequencies. The energy, $E=h\nu$, of each photon is multiplied by the number of states available at that frequency, and the probability that each of those states will be occupied.

Integrating the above equation over ν the power output given by the Stefan–Boltzmann law is obtained, as:

$$P=\sigma\cdot A\cdot T^4$$

where the constant of proportionality σ is the Stefan–Boltzmann constant and A is the radiating surface area.

The wavelength λ, for which the emission intensity is highest, is given by Wien's displacement law as:

$$\lambda_{max}=\frac{b}{T}$$

For surfaces which are not black bodies, one has to consider the (generally frequency dependent) emissivity factor $\epsilon(\nu)$. This factor has to be multiplied with the radiation spectrum formula before integration. If it is taken as a constant, the resulting formula for the power output can be written in a way that contains ϵ as a factor:

$$P=\epsilon\cdot\sigma\cdot A\cdot T^4$$

This type of theoretical model, with frequency-independent emissivity lower than that of a perfect black body, is often known as a *grey body*. For frequency-dependent emissivity, the solution for the integrated power depends on the functional form of the dependence, though in general there is no simple expression for it. Practically speaking, if the emissivity of the body is roughly constant around the peak emission wavelength, the gray body model tends to work fairly well since the weight of the curve around the peak emission tends to dominate the integral.

Constants

Definitions of constants used in the above equations:

h	Planck's constant	$6.626\ 069\ 3(11)\times10^{-34}$ J·s = $4.135\ 667\ 43(35)\times10^{-15}$ eV·s
b	Wien's displacement constant	$2.897\ 768\ 5(51)\times10^{-3}$ m·K
k_B	Boltzmann constant	$1.380\ 650\ 5(24)\times10^{-23}$ J·K^{-1} = $8.617\ 343\ (15)\times10^{-5}$ eV·K^{-1}
σ	Stefan–Boltzmann constant	$5.670\ 373\ (21)\times10^{-8}$ W·m^{-2}·K^{-4}
c	Speed of light	299 792 458 m·s^{-1}

Variables

Definitions of variables, with example values:

T	Absolute temperature	For units used above, must be in kelvins (e.g. average surface temperature on Earth = 288 K)
A	Surface area	$A_{cuboid} = 2ab + 2bc + 2ac$; $A_{cylinder} = 2\pi \cdot r(h + r)$; $A_{sphere} = 4\pi \cdot r^2$

Radiative Heat Transfer

The *net* radiative heat transfer from one surface to another is the radiation leaving the first surface for the other minus that arriving from the second surface.

- For black bodies, the rate of energy transfer from surface 1 to surface 2 is:

$$\dot{Q}_{1\to2} = A_1 E_{b1} F_{1\to2} - A_2 E_{b2} F_{2\to1}$$

where A is surface area, E_b is energy flux (the rate of emission per unit surface area) and $F_{1\to2}$ is the view factor from surface 1 to surface 2. Applying both the reciprocity rule for view factors, $A_1 F_{1\to2} = A_2 F_{2\to1}$, and the Stefan–Boltzmann law, $E_b = \sigma T^4$, yields:

$$\dot{Q}_{1\to2} = \sigma A_1 F_{1\to2}(T_1^4 - T_2^4)$$

where σ is the Stefan–Boltzmann constant and T is temperature. A negative value for $\dot{Q}$ indicates that net radiation heat transfer is from surface 2 to surface 1.

- For two grey-body surfaces forming an enclosure, the heat transfer rate is:

$$\dot{Q} = \frac{\sigma(T_1^4 - T_2^4)}{\frac{1-\epsilon_1}{A_1\epsilon_1} + \frac{1}{A_1 F_{1\to2}} + \frac{1-\epsilon_2}{A_2\epsilon_2}}$$

where ϵ_1 and ϵ_2 are the emissivities of the surfaces.

Formulas for radiative heat transfer can be derived for more particular or more elaborate physical arrangements, such as between parallel plates, concentric spheres and the internal surfaces of a cylinder.

Kirchhoff's Law of Thermal Radiation

In general, both the emissivity, ε, and the absorptivity, α, of a surface depend on the temperature and the wavelength of the radiation. Kirchhoff's law of thermal radiation, postulated by a German physicist Gustav Robert Kirchhoff, states that the emissivity and the absorptivity of a surface at a given temperature and wavelength are equal.

For an arbitrary body emitting and absorbing thermal radiation in thermodynamic equilibrium, the emissivity is equal to the absorptivity.

emissivity ε = absorptivity α

Proof:

Let e_λ and a_λ be the monochromatic emittance and absorptance of the body for wavelength at temperature T.

Let Q be the amount of radiation incident per unit area on the body in the wavelength range (λ–1/2) to (λ + 1/2).

Therefore,

Amount of radiation absorbed by body per second per unit area = a Q.

The body is in thermal equilibrium with the enclosure.

Therefore, the amount of radiations absorbed by body per sec per unit area in unit wavelength region is equal to the amount of radiations emitted by body per sec per unit area in the same unit wavelength region.

$$a_\lambda Q = e_\lambda$$

i.e.

$$Q = \frac{e_\lambda}{a_\lambda}$$

Therefore, for black body

$$a_\lambda = 1$$

$$\therefore \frac{E_\lambda}{1} = Q$$

where,

E_λ is the emissive power of black body.

From above equation, we have,

$\frac{e_\lambda}{\alpha_\lambda} = E_\lambda = \text{constant}$, which is Kirchhoff's law.

Also from above equation, we have,

$$e_\lambda = E_\lambda \, a_\lambda$$
$$\Rightarrow e_\lambda \, \alpha \, a_\lambda$$

i.e. good absorbers of radiation are also good emitters of radiation.

This law must be also valid in order to satisfy the Second Law of Thermodynamics. As was written, all bodies above absolute zero temperature radiate some heat. Two objects radiate heat toward each other. But what if a colder object with high emissivity radiates toward a hotter object with very low emissivity? This seems to violate the Second Law of Thermodynamics, which states that heat cannot spontaneously flow from cold system to hot system without external work being performed on the system. The paradox is resolved by the fact that each body must be in direct line of sight of the other to receive radiation from it. Therefore, whenever the cool body is radiating heat to the hot body, the hot body must also be radiating heat to the cool body. Moreover, the hot body will radiate more energy than cold body. The case of different emissivities is solved by the Kirchhoff's Law of thermal radiation, which states that object with low emissivity have also low absorptivity. As a result, heat cannot spontaneously flow from cold system to hot system and the second law is still satisfied.

References

- Chen, G. (2004). Nanoscale energy transport and conversion : a parallel treatment of electrones, molecules, phonons, and photons. New York: Oxford. ISBN 978-0195159424
- Onsager, L. (1931). "Reciprocal Relations in Irreversible Processes. I.". Physical Review. 37 (4): 405–426. Bibcode:1931PhRv...37..405O. doi:10.1103/PhysRev.37.405
- Blackmore, Carolyn. "Classification and Handling of PCB Waste" (PDF). Lawrence Berkeley National Laboratory. Retrieved 20 October 2017
- Bahiraei, Mehdi (2016-11-01). "Particle migration in nanofluids: A critical review". International Journal of Thermal Sciences. 109: 90–113. doi:10.1016/j.ijthermalsci.2016.05.033
- Edited by Kanatzidis, M.G.;Mahanti, S. D.;Hogan, T. P. (2003). Chemistry, physics, and materials science of thermoelectric materials : beyond bismuth telluride. New York [u.a.]: Kluwer Academic/Plenum Publ. ISBN 0306477386
- Carey, V. P.; Chen, G.; Grigoropoulos, C.; Kaviany, M.; Majumdar, A. (2008). "A Review of Heat Transfer Physics". Nanoscale and Microscale Thermophysical Engineering. 12 (1): 1–60. doi:10.1080/15567260801917520
- Buongiorno, J. (March 2006). "Convective Transport in Nanofluids". Journal of Heat Transfer. American Society Of Mechanical Engineers. 128 (3): 240. doi:10.1115/1.2150834. Retrieved 27 March 2010

- Pak, Choi (23 April 2007). "Hydrodynamic And Heat Transfer Study of Dispersed Fluids With Submicron Metallic Oxide Particles". A Journal of Thermal Energy
- Gill, Paul (2009). Electrical power equipment maintenance and testing (2nd ed.). Boca Raton: CRC Press. p. 193. ISBN 978-1-57444-656-2
- Shin, S.; Kaviany, M. (2009). "Enhanced laser cooling of CO_2–Xe gas using (02°0) excitation". Journal of Applied Physics. 106 (12): 124910. Bibcode:2009JAP...106l4910S. doi:10.1063/1.3273488
- Das, Sarit K.; Stephen U. S. Choi; Wenhua Yu; T. Pradeep (2007). Nanofluids: Science and Technology. Wiley-Interscience. p. 397. Retrieved 27 March 2010
- "Conceptions for heat transfer correlation of nanofluids". International Journal of Heat and Mass Transfer. 43 (19): 3701–3707. 2000-10-01. doi:10.1016/S0017-9310(99)00369-5. ISSN 0017-9310
- Sakurai, J.J. (1973). Advanced quantum mechanics (4th printing, with revisions. ed.). Menlo Park, California: Benjamin/Cummings. ISBN 0201067102
- Djeu, N.; Whitney, W. (1981). "Laser Cooling by Spontaneous Anti-Stokes Scattering". Physical Review Letters. 46 (4): 236–239. Bibcode:1981PhRvL..46..236D. doi:10.1103/PhysRevLett.46.236
- "Enhancing the efficiency of polymerase chain reaction using graphene nanoflakes - Abstract - Nanotechnology - IOPscience". iop.org. Retrieved 8 June 2015
- "Nanofluid for enhanced oil recovery". Journal of Petroleum Science and Engineering. 78 (2): 431–437. 2011-08-01. doi:10.1016/j.petrol.2011.06.014. ISSN 0920-4105

Heat Conduction

Heat is transferred within a body due to microscopic collisions between particles and electrons. This is termed as heat conduction. All the diverse aspects of heat conduction, such as heat equation, Fourier's law for heat conduction, thermal diffusivity, thermal conductivity, thermal effusivity, etc. have been carefully analyzed in this chapter.

Heat can only be transferred through three means: conduction, convection and radiation. Of these, conduction is perhaps the most common, and occurs regularly in nature. In short, it is the transfer of heat through physical contact. It occurs when you press your hand onto a window pane, when you place a pot of water on an active element, and when you place an iron in the fire.

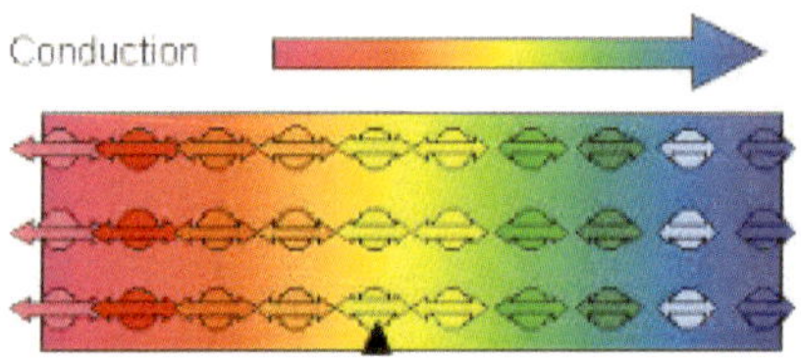

This transfer occurs at the molecular level—from one body to another—when heat energy is absorbed by a surface and causes the molecules of that surface to move more quickly. In the process, they bump into their neighbors and transfer the energy to them, a process which continues as long as heat is still being added.

The process of heat conduction depends on four basic factors: the temperature gradient, the cross section of the materials involved, their path length, and the properties of those materials.

A temperature gradient is a physical quantity that describes in which direction and at what rate the temperature changes in a specific location. Temperature always flows from the hottest to coldest source, due to the fact that cold is nothing but the absence of heat energy. This transfer between bodies continues until the temperature difference decays, and a state known as thermal equilibrium occurs.

Cross-section and path length are also important factors. The greater the size of the material involved in the transfer, the more heat is needed to warm it. Also, the more surface area that is exposed to open air, the greater likelihood for heat loss. So shorter objects with a smaller cross-section are the best means of minimizing the loss of heat energy.

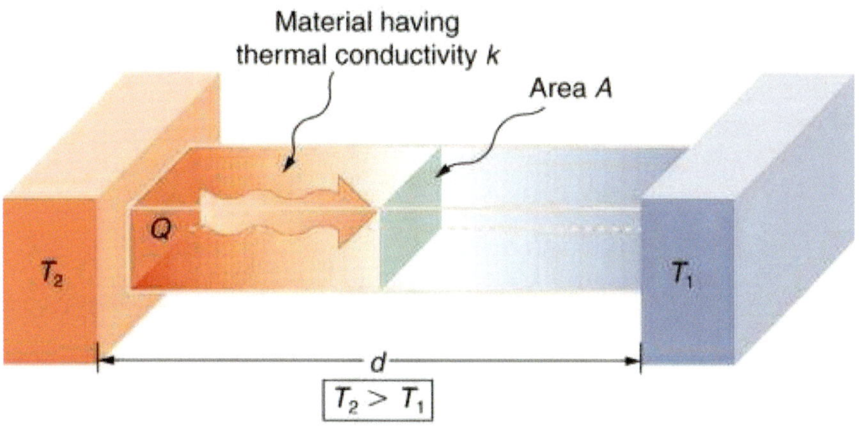

Heat conduction occurs through any material, represented here by a rectangular bar. The rate at which it is transfers depends in part on the thickness of the material (rep. by A)

Last, but certainly not least, is the physical properties of the materials involved. Basically, when it comes to conducting heat, not all substances are created equal. Metals and stone are considered good conductors since they can speedily transfer heat, whereas materials like wood, paper, air, and cloth are poor conductors of heat.

These conductive properties are rated based on a "coefficient" which is measured relative to silver. In this respect, silver has a coefficient of heat conduction of 100, whereas other materials are ranked lower. These include copper (92), iron (11), water (0.12), and wood (0.03). At the opposite end of the spectrum is a perfect vacuum, which is incapable of conducting heat, and is therefore ranked at zero.

Materials that are poor conductors of heat are called insulators. Air, which has a conduction coefficient of.006, is an exceptional insulator because it is capable of being contained within an enclosed space. This is why artificial insulators make use of air compartments, such as double-pane glass windows which are used for cutting heating bills. Basically, they act as buffers against heat loss.

Feather, fur, and natural fibers are all examples of natural insulators. These are materials that allows birds, mammals and human beings to stay warm. Sea otters, for example, live in ocean waters that are often very cold and their luxuriously thick fur keeps them warm. Other sea mammals like sea lions, whales and penguins rely on thick layers of fat (aka. blubber) – a very poor conductor – to prevent heat loss through their skin.

This same logic is applied to insulating homes, buildings, and even spacecraft. In these cases, methods involve either trapped air pockets between walls, fiber-glass (which traps air within it) or high-density foam. Spacecraft are a special case, and use insulation in the form of foam, reinforced carbon composite material, and silica fiber tiles. All of these are poor conductors of heat, and therefore prevent heat from being lost in space and also prevent the extreme temperatures caused by atmospheric reentry from entering the crew cabin.

Conduction, as demonstrated by heating a metal rod with a flame

The laws governing conduction of heat are very similar to Ohm's Law, which governs electrical conduction. In this case, a good conductor is a material that allows electrical current (i.e. electrons) to pass through it without much trouble. An electric insulator, by contrast, is any material whose internal electric charges do not flow freely, and therefore make it very hard to conduct an electric current under the influence of an electric field.

In most cases, materials that are poor conductors of heat are also poor conductors of electricity. For instance, copper is good at conducting both heat and electricity, hence why copper wires are used so widely in the manufacture of electronics. Gold and silver are even better, and where price is not an issue, these materials are used in the construction of electrical circuits as well.

And when one is looking to "ground" a charge (i.e. neutralize it), they send it through a physical connection to the Earth, where the charge is lost. This is common with electrical circuits where exposed metal is a factor, ensuring that people who accidentally come into contact are not electrocuted.

This view of the nose section of space shuttle Discovery, build of heat-resistance carbon-composites. Credit: NASA

Insulating materials, such as rubber on the soles of shoes, is worn to ensure that people working with sensitive materials or around electrical sources are protected from electrical charges. Other insulating materials like glass, polymers, or porcelain are commonly used on power lines and high-voltage power transmitters to keep power flowing to the circuits.

For heat transfer between two plane surfaces, such as heat loss through the wall of a house, the rate of conduction heat transfer is:

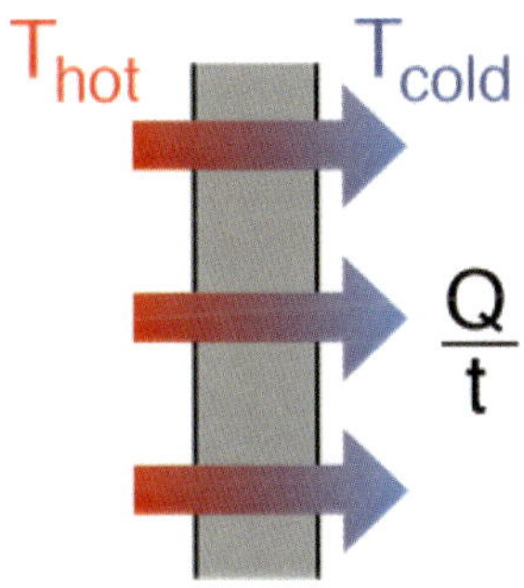

$$\frac{Q}{t} = \frac{kA\left(T_{hot} - T_{cold}\right)}{d}$$

Q = heat transferred in time = t

k = thermal conductivity of the barrier

A = area

T = temperature

d = thickness of barrier

Relativistic Heat Conduction

Relativistic heat conduction refers to the modelling of heat conduction (and similar diffusion processes) in a way not compatible with special relativity. It discusses models using a wave equation with a dissipative term.

Ali and Zhang claim their model of relativistic heat conduction is the only one compatible with the theory of special relativity, the second law of thermodynamics, electrodynamics, and quantum mechanics, simultaneously. The main features of their model are:

1. It admits a finite speed of heat propagation, and allows for relativistic effects when heat flux transients approach that speed.
2. It removes the possibility of paradoxical situations that may violate the second law of thermodynamics.
3. It, implicitly, admits the wave–particle duality of the heat-carrying "phonon".

These outcomes are achieved by (1) upgrading the Fourier equation of heat conduction to the form of a Telegraph equation of electrodynamics, and (2) introducing a new definition of the heat flux vector. Consequently, their model gives rise to a number of phenomena, such as thermal resonance and thermal shock waves, which are possible during high-frequency pulsed laser heating of thermal insulators.

Background

Classical Model

For most of the last two centuries, heat conduction has been modelled by the well-known Fourier equation:

$$\frac{\partial \theta}{\partial t} = \alpha \nabla^2 \theta,$$

where θ is temperature, t is time, $\alpha = k/(\rho\, c)$ is thermal diffusivity, k is thermal conductivity, ρ is density, and c is specific heat capacity. The Laplace operator, ∇^2, is defined in Cartesian coordinates as

$$\nabla^2 = \frac{\partial^2}{\partial x^2} + \frac{\partial^2}{\partial y^2} + \frac{\partial^2}{\partial z^2}.$$

This Fourier equation can be derived by substituting Fourier's linear approximation of the heat flux vector, q, as a function of temperature gradient,

$$\mathbf{q} = -k \nabla \theta,$$

into the first law of thermodynamics

$$\rho c \frac{\partial \theta}{\partial t} + \nabla \cdot \mathbf{q} = 0,$$

where the del operator, ∇, is defined in 3D as

$$\nabla = \frac{\partial}{\partial x}\mathbf{i} + \frac{\partial}{\partial y}\mathbf{j} + \frac{\partial}{\partial z}\mathbf{k}.$$

It can be shown that this definition of the heat flux vector also satisfies the second law of thermodynamics,

$$\nabla \cdot \left(\frac{\mathbf{q}}{\theta}\right) + \rho \frac{\partial s}{\partial t} = \sigma,$$

where s is specific entropy and σ is entropy production. Alternatively, the second law can be written as

$$\sigma = \frac{-1}{\theta^2} \mathbf{q} \cdot \nabla \theta,$$

which leads to the condition

$$\sigma=\frac{k}{\theta^2}\left[\left(\frac{\partial\theta}{\partial x}\right)^2+\left(\frac{\partial\theta}{\partial y}\right)^2+\left(\frac{\partial\theta}{\partial z}\right)^2\right],$$

which is always true, because k is a non-negative material property.

Hyperbolic Model

It is well known that the Fourier equation (and the more general Fick's law of diffusion) is incompatible with the theory of relativity for at least one reason: it admits infinite speed of propagation of heat signals within the continuum field. For example, consider a pulse of heat at the origin; then according to Fourier equation, it is felt (i.e. temperature changes) at any distant point, instantaneously. The speed of information propagation is faster than the speed of light in vacuum, which is inadmissible within the framework of relativity.

To overcome this contradiction, workers such as Cattaneo, Vernotte, Chester, and others proposed that Fourier equation should be upgraded from the parabolic to a hyperbolic form,

$$\frac{1}{C^2}\frac{\partial^2\theta}{\partial t^2}+\frac{1}{\alpha}\frac{\partial\theta}{\partial t}=\nabla^2\theta\,.$$

In this equation, C is called the speed of second sound (i.e. the fictitious quantum particles, phonons). The equation is known as the hyperbolic heat conduction (HHC) equation. Mathematically, it is the same as the telegrapher's equation, which is derived from Maxwell's equations of electrodynamics.

For the HHC equation to remain compatible with the first law of thermodynamics, it is necessary to modify the definition of heat flux vector, q, to

$$\tau_0\frac{\partial\mathbf{q}}{\partial t}+\mathbf{q}=-k\,\nabla\theta,$$

where τ_0 is a relaxation time, such that

The most important implication of the hyperbolic equation is that by switching from a parabolic (dissipative) to a hyperbolic (includes a conservative term) partial differential equation, there is the possibility of phenomena such as thermal resonance and thermal shock waves.

Criticism of the HHC Model

- The relaxation time, τ_0, is justified based on microscopic aspects of lattice vibration and electron transport; is an extension of kinetic theory calculations and Boltzmann equation for rarefied gases to the case of solids; and is calculated from statistical Newtonian mechanics. Further, the speed C is only a collection of terms, α and τ_0, and has no physical reality or significance similar to that associated with the speed of light. Hence, the hyperbolic equation is compatible with relativity artificially (in form only), but is still fundamentally classical Newtonian.

- The new definition of heat flux vector is an *ad hoc* mathematical approximation of a far more complicated expression; this raises some doubts about the whole approach.
- The most serious criticism is that the hyperbolic equation can violate the second law of thermodynamics. For example, consider an infinitely long wire conductor, with a heat source at the origin, and measure temperature at distances significantly remote from origin. If the heat source at origin varies with a frequency much higher than the relaxation time (i.e. faster than the speed of second sound) then the hyperbolic equation admits a temperature field in which heat would appear to be moving from cold to hot, in violation of the second law. This contradiction was demonstrated in more mathematically rigorous form.

Ali and Zhang's theory of RHC attempts to resolve the controversies surrounding the hyperbolic equation, while maintaining the form of that equation. This is achieved by:

- Deriving the hyperbolic equation starting from space–time duality of a Minkowski space, and simple Lorentz transformations, that are basic to the theory of special relativity. This is done without any reference to microstructure or statistical mechanics.
- Treating the speed of second sound, C, as a fundamental property of the temperature field, although still fundamentally inferior to the speed of light.
- Modifying the definition of the heat flux vector so that it is simpler, more elegant, and bring it in compliance with the second law of thermodynamics.

Derivation of the RHC Equation

Transformations

In a Euclidean space, distance between any two points, ds, is measured by

$$ds^2 = dx^2 + dy^2 + dz^2,$$

where dx, dy, and dz are displacements along three orthogonal axes.

In a Minkowski space, distance between two events, ds, is measured by

$$ds^2 = d\tau^2 + dx^2 + dy^2 + dz^2,$$

where, τ, is space-like-time and is related to real time, t, by

$$\tau = iCt,$$

where C is speed of light in vacuum and $i^2 = -1$. Hence,

$$ds^2 = dx^2 + dy^2 + dz^2 - C^2\, dt^2.$$

Consequently, the 3D del, ∇,operator is upgraded to the 4D quad, $\square$, operator (also known as the Four-gradient)

$$\square = \frac{\partial}{\partial\tau}\mathbf{o} + \frac{\partial}{\partial x}\mathbf{i} + \frac{\partial}{\partial y}\mathbf{j} + \frac{\partial}{\partial z}\mathbf{k} = \frac{\partial}{\partial\tau}\mathbf{o} + \nabla = \frac{-i}{C}\frac{\partial}{\partial t}\mathbf{o} + \nabla.$$

Likewise, the 3D Laplacian, ∇^2, operator is upgraded to the 4D d'Alembert operator,

$$\square^2 = \frac{\partial^2}{\partial\tau^2} + \frac{\partial^2}{\partial x^2} + \frac{\partial^2}{\partial y^2} + \frac{\partial^2}{\partial z^2} = \frac{\partial^2}{\partial\tau^2} + \nabla^2 = \frac{-1}{C^2}\frac{\partial^2}{\partial t^2} + \nabla^2.$$

Any physical quantity that is Galilean invariant in Euclidean space can be made Lorentz invariant in a Minkowski space, by upgrading from 3D to 4D operators. Consequently, Fourier's equation can be upgraded to 4D as

$$\frac{\partial\theta}{\partial t} = \alpha\square^2\theta = \frac{-\alpha}{C^2}\frac{\partial^2\theta}{\partial t^2} + \alpha\nabla^2\theta,$$

which is called the relativistic heat conduction equation. Likewise, the definition of the heat-flux vector, q, is upgraded to the 4D form as

$$\mathbf{q} = -k\square\theta = -k\nabla\theta + \frac{ik}{C}\frac{\partial\theta}{\partial t}\mathbf{o}.$$

Implications

It can be shown that this definition of q is compatible with the first law of thermodynamics,

$$\rho c\frac{\partial\theta}{\partial t} + \square\cdot\mathbf{q} = 0 = \rho c\frac{\partial\theta}{\partial t} + \nabla\cdot\mathbf{q} + \frac{-i}{C}\frac{\partial\mathbf{q}}{\partial t}\cdot\mathbf{o},$$

as well as the second law of thermodynamics,

$$\square\cdot\left(\frac{\mathbf{q}}{\theta}\right) + \rho\frac{\partial s}{\partial t} = \sigma,$$

in their 4D upgraded form. The imaginary terms in these equations are direct manifestation of the wave nature of heat, and are essential for the heat equation to become compatible with all laws of physics. The real terms in these equations are identical to those in the classical heat model.

It has been observed that RHC reduces the second law of thermodynamics to a statement of the form

$$\left(\frac{dt}{dx}\right)^2+\left(\frac{dt}{dy}\right)^2+\left(\frac{dt}{dz}\right)^2\geqslant\frac{1}{C^2},$$

which is the "no action at a distance" principle of special relativity. Essentially, the RHC asserts that relativity and the second law of thermodynamics are two alternative, but equal statements about the nature of time. Both physical principles are mutually derivable from each other and are complementary.

As far as heat conduction is concerned, the RHC equation is identical in form to the hyperbolic equation, and all analytical and experimental results that are relevant to one are equally applicable to the other. The definition of heat flux vector, however, is different; but the RHC definition is merely a 4D upgrade of the original linear Fourier approximation. The mathematics of RHC is much simpler and more elegant. However, RHC raises some significant conceptual challenges:

1. This weak interpretation of relativity, in which the speed of second sound plays a role similar to that of the speed of light, can be viewed as downgrading or degrading to the universality of the theory of relativity. Notice how the symbol c in standard relativity theory is replaced with C without much interpretation.

2. The implied wave nature of heat is controversial. Some workers reject the wave nature of heat on dogmatic grounds. Moreover, RHC implies that a phonon is a full-fledged objective quantum particle whose physical reality is no lesser than that of a photon. Existing experimental evidences are not enough to support for or against such views.

3. Heat quantities become complex numbers, with values including "imaginary temperature", which are cannot be interpreted experimentally.

4. The equivalence of relativity and the second law is shocking, because it implies that one of them can be a derivative of the other.

In summary, while the RHC is mathematically simple and elegant, and experimentally practical and relevant, it raises a number of conceptual issues that are highly controversial.

Applications

The RHC theory is applicable for any physical problem in which the hyperbolic equation is relevant: when speed of heat propagation is small, e.g. thermal insulators, or when speed of heat-flux variation is very large, e.g. pulsed-laser heating. Applications for those types of problems are abundant, and there is plenty of published work. Most of these results remain relevant to RHC, but because the definition of heat flux vector is different, final closed-form solutions may not be the same. In many cases, RHC provides closed-form solutions that are not possible using the HHC model. A number of useful fundamental solutions for 1D and 2D relativistic moving heat sources are available in closed-form.

Thermal Contact Conductance

In physics, thermal contact conductance is the study of heat conduction between solid bodies in thermal contact. The thermal contact conductance coefficient, h_c, is a property indicating the thermal conductivity, or ability to conduct heat, between two bodies in contact. The inverse of this property is termed thermal contact resistance.

The temperature distribution near an interface between two solid bodies in contact when heat flows normally from one body to the other. Interest in accurate understanding of this has increased recently, due to the need to know as accurately as possible the temperature in fuel elements of nuclear reactors and other equipment in which high heat fluxes flow from one body to another. Much of the experimental and theoretical work in this field has been done in the last fifteen years, though mathematical groundwork was laid in the 19th century.

Two bodies which are in contact at nominally flat surfaces will actually touch only at a few discrete spots. Hence the exact temperature distribution is complicated and threedimensional. An approximation, adequate for many purposes, is to define and determine a hypothetical temperature drop $\Delta T_C = (T_1 - T_2)$. At a distance from the interface which is large compared with the typical spacing between contacts the temperature in each body is then taken to be:

$$T_1\ (or\ T_2) = \frac{\text{heat flux x distance from interface}}{\text{conductivity}}$$

Provided the relative positions of the bodies do not change while the heat flux is varied, it is found that ΔT, is proportional to the heat flux (Q/A), and the constant of proportionality, defined by $h_c = (Q/A)/\Delta T$, is known as the thermal interfacial contact conductance, abbreviated herein to contact conductance.

Single Contacts

Useful theoretical studies of the flow of heat between solid bodies have been based on considering the flow of heat through one microscopic contact region and adjacent solid. This problem, of flow of heat through abutting cylinders, was studied by Cetinkale and Fishenden who showed that, even if the cylinders had different conductivity and heat is also conducted through the interstitial gap, there exists an isothermal plane at z = 0 (suitably subdividing the interstitial gap). The problem involving two cylinders can thus be reduced to the simpler problem of one cylinder, with in the present case, temperature specified *over* part of the boundary and heat flux specified over the remainder. Although simpler, this problem has nevertheless defied exact analytic solution, due to the mixed boundary conditions, though an exact numerical solution has recently been reported by Clausing. Some early work on the analogous electrical problem contained approximations which were only close for very small values of (c/b). Solutions for larger (c/b) were obtained independently by Roess and by Mikic, who replaced the temperature boundary c5ndition at z = 0 by a distribution of heat flux:

$$\textit{proportional to} \frac{1}{(c^2 - r^2)^{\frac{1}{2}}} \quad \text{for } r < c$$

$$\text{zero} \quad \text{for } c < r < b.$$

This distribution was chosen so as to make T nearly constant at z = 0, r < c. The revised problem could be solved exactly. This has apparently not been described outside the report literature.

$$h_c = 2k\frac{c}{A_0} \times \frac{1}{\psi}$$

where

$$\frac{1}{k} = \frac{1}{2}\left(\frac{1}{k_1} + \frac{1}{k_2}\right)$$

and

ψ is given by equation. The results obtained by Roess and by Mikic, though expressed in different algebraic form, agree closely for 0 < (c/b) < 0.4.

Clausing reports a numerical calculation of the heat flow pattern using the true boundary conditions of zero flow at r = b and at z = 0, $r > c$ and constant temperature at z = 0, $r < c$. His result and those of Roess and Mikic are in close agreement for 0 < (c/b) < 0-4 and only differ by a few percent for (c/b) = 0.6.

Clausing's is the only calculation which aims at high accuracy for large (c/b), so the graph of ψ has been continued beyond *(c/b)* = 0.4 by using table as far as that table goes, i.e. to (c/b) = 0.833. Contacts with still larger values of (c/b) will have small resistance, so the precise value of ψ for larger *(c/b)* is of little practical importance.

As indicated in figure below, various approximations can be made for ψ and the choice between them will depend on the range of *(c/b)* involved and the accuracy needed.

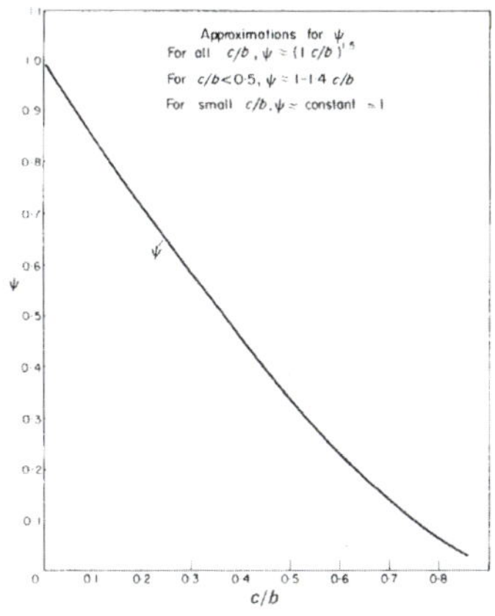

Contact resistance function ψ

Multiple Contacts

In the previous section the contact conductance was derived for a single circular contact spot placed centrally in a cylindrical region. The aim of this section is to extend that work towards more practical cases by considering m contacts of areas A_{r1}, A_{r2}, ... A_{rm} (of equivalent radii c_1, c_2,.. c_m,) distributed over an area of apparent contact A_a, between abutting cylinders. It is still assumed that the interface is nearly plane, and that the heat enters and leaves the cylinders over end surfaces at great distance from the interface.

Equality of Temperatures at Contacts

On these assumptions the local configuration Body 2 is as shown in figure, and it can be shown that all contacts must be at the same temperature, provided in each body the conductivity is independent of direction, position and temperature. The proof can be based on the more Fig. Two bodies having mirror symmetry about plane general symmetrical case shown in Fig., where

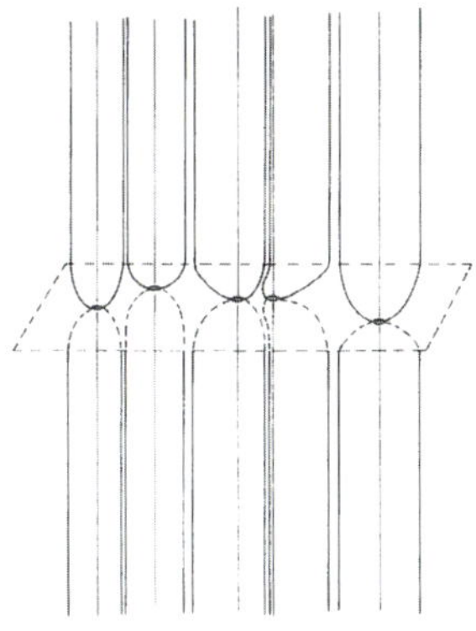

General configuration and heat flow for multiple contacts at plane interface

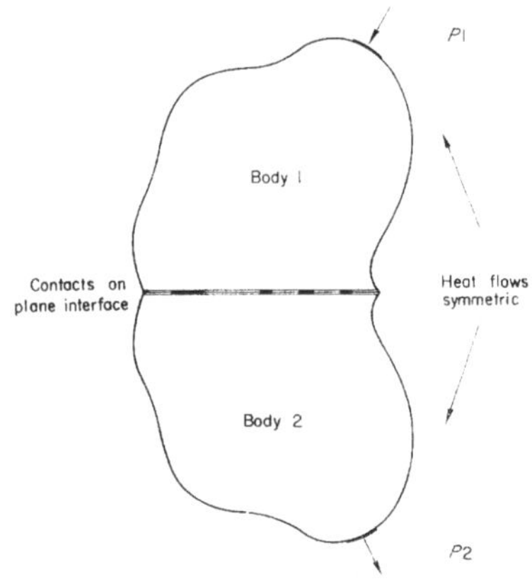

Two bodies having mirror symmetry about plane interface

there is mirror symmetry in the sense that if $P_1(x_1, y_1, z_1)$ is on the surface of body 1, then $P_2(x_2, y_2, -z_1)$ is on the surface of body 2, and in addition, for any typical specified inflow of heat into body 1 at point P_1 there is a corresponding outflow from body 2 at P_2. For this case we need a solution of:

$$\nabla^2 T = 0 \text{ in each body}$$

with :

$$k\frac{\partial T}{\partial n} = \text{specified values at points on boundary,}$$

and at the interface z = 0 :

$$k_1\frac{\partial T_1}{\partial z} = \left(T_1 - T_2\right)h_v = k_2\frac{\partial T_2}{\partial z} \text{ outside contact spots}$$

$$k_1\frac{\partial T_1}{\partial z} = k_2\frac{\partial T_2}{\partial z}, \qquad T_1 = T_2 \text{ inside contact spots}$$

where h_r is a void conductance, not necessarily constant.

This problem has a unique solution, apart from an additive constant, so the problem is virtually solved if we can construct one solution, by any means.

A solution can be constructed by solving for body 1:

$$\nabla^2 (k_1 T) = 0$$

With

$$\frac{\partial\left(k_1 T\right)}{\partial n_1} = \text{specified values at P, etc.}$$

and

$$\left.\begin{array}{ll} \dfrac{\partial(k_1T)}{\partial z}=h_v\left\{\dfrac{1}{k_1}+\dfrac{1}{k_2}\right\}(k_1T) & outside\ contact\ spots \\ k_1T=0 & inside\ contact\ spots \end{array}\right\}\text{at } z=0$$

and solving for body 2 similar equations with suffices 1, 2 interchanged, the direction of the normal outward instead ofinward and a negative sign before h_v.

The two sets of equations are identical in terms of *(kT)* apart from sign changes which may be assigned to *T* and z. Hence the two solutions are also identical except for these sign changes. The solutions therefore have negative mirror symmetry in the sense that the value of (k_1T_1) at (x_0, y_0, z_0) is equal to minus the value of (k_2T_2) at $(x_0, y_0, -z_0)$. It follows that the various boundary conditions at z = 0 are also all met, and by taking T_1 in body 1 and in body 2 we have a solution for the complete problem shown in Fig. Since that problem has a unique solution, this must be the solution (apart from additive constant). Hence the contacts are all at the same temperature for the case shown in Fig., and for the special case shown in Fig.

Provided $h_v = 0$, the condition requiring conductivity to be independent of temperature can be relaxed by use of the substitution $T' = (1/k_0)\int k\,\mathrm{d}T$. Symmetry will then be in terms of (k_0T').

Definition

According to Fourier's law, the heat flow between the bodies is found by the relation:

$$q = -kA\frac{dT}{dx}$$

where q is the heat flow, k is the thermal conductivity, A is the cross sectional area and dT/dx is the temperature gradient in the direction of flow.

From considerations of energy conservation, the heat flow between the two bodies in contact, bodies A and B, is found as:

$$q = \frac{T_1 - T_3}{\Delta x_A/(k_AA) + 1/(h_cA) + \Delta x_B/(k_BA)}$$

One may observe that the heat flow is directly related to the thermal conductivities of the bodies in contact, k_A and k_B, the contact area A, and the thermal contact resistance, $1/h_c$, which, as previously noted, is the inverse of the thermal conductance coefficient, h_c.

Importance

Most experimentally determined values of the thermal contact resistance fall between 0.000005 and 0.0005 m^2 K/W (the corresponding range of thermal contact conductance is 200,000 to 2000 W/m^2 K). To know whether the thermal contact resistance is significant or not, magnitudes of the thermal resistances of the layers are compared with typical values of thermal contact resistance. Thermal contact resistance is significant and may dominate for good heat conductors such as metals but can be neglected for poor heat conductors such as insulators. Thermal contact conductance is an important factor in a variety of applications, largely because many physical systems contain a mechanical combination of two materials. Some of the fields where contact conductance is of importance are:

- Electronics
  - Electronic packaging
  - Heat sinks
  - Brackets
- Industry
  - Nuclear reactor cooling
  - Gas turbine cooling
  - Internal combustion engines
  - Heat exchangers
  - Thermal insulation
- Flight
  - Hypersonic flight vehicles
  - Thermal supervision for space vehicles
- Residential/Building Science
  - Performance of building envelops

Factors Influencing Contact Conductance

Thermal contact conductance is a complicated phenomenon, influenced by many factors. Experience shows that the most important ones are as follows:

An enlargement of the interface between two contacting surfaces.
The finish quality is exaggerated for the sake of the argument

Contact Pressure

The contact pressure is the factor of most influence on contact conductance. As contact pressure grows, contact conductance grows (And consequentially, contact resistance becomes smaller). This is attributed to the fact that the contact surface between the bodies grows as the contact pressure grows.

Since the contact pressure is the most important factor, most studies, correlations and mathematical models for measurement of contact conductance are done as a function of this factor.

The thermal contact resistance of certain sandwich kinds of materials that are manufactured by rolling under high temperatures may sometimes be ignored because the decrease in thermal conductivity between them is negligible.

Interstitial Materials

No truly smooth surfaces really exist, and surface imperfections are visible under a microscope. As a result, when two bodies are pressed together, contact is only performed in a finite number of points, separated by relatively large gaps, as can be shown in figure above. Since the actual contact area is reduced, another resistance for heat flow exists. The gases/fluids filling these gaps may largely influence the total heat flow across the interface. The thermal conductivity of the interstitial material and its pressure are the two properties governing its influence on contact conductance.

In the absence of interstitial materials, as in a vacuum, the contact resistance will be much larger, since flow through the intimate contact points is dominant.

Surface Roughness, Waviness and Flatness

One can characterise a surface that has undergone certain finishing operations by three properties: roughness, waviness and flatness. Among these, roughness is of most importance, and is usually indicated by an rms value, σ.

Surface Deformations

When the two bodies come in contact, surface deformation may occur on both bodies. This deformation may either be plastic or elastic, depending on the material properties and the contact pressure. When a surface undergoes plastic deformation, contact resistance is lowered, since the deformation causes the actual contact area to increase

Surface Cleanliness

The presence of dust particles, acids, etc., can also influence the contact conductance.

Thermal Boundary Conductance

While a finite thermal contact conductance is due to voids at the interface, surface waviness, and surface roughness, etc., a finite conductance exists even at near ideal interfaces as well. This conductance, known as thermal boundary conductance, is due to the differences in electronic and vibrational properties between the contacting materials. This conductance is generally much higher than thermal contact conductance, but becomes important in nanoscale material systems.

Heat Equation

The heat equation is a differential equation involving three variables – two independent variables x and t, and one dependent variable u = u(t, x).

The equation states:

$$\frac{\partial u}{\partial t} = k \frac{\partial^2 u}{(\partial x)^2}$$

$k \in \mathbb{R}$ is a real number.

Here the symbol $\partial u / \partial t$ means the derivative of u with respect to t, keeping x constant, while the symbol $\partial^2 u / (\partial x)^2$ denotes the second partial derivative in x, keeping t constant.

Terminology Associated with the Heat Equation

For the heat equation, we use the following terminology (which shall be clear once we get to the physical motivation):

1. The variable t is termed the time parameter, or the time variable.

2. The variable x is termed the spatial parameter, or the spatial variable.
3. The function u is termed the heat function

The n-Dimensional Heat Equation

The heat equation in n dimensions is defined as follows:

There are n + 1 independent variables, namely t (the time parameter) and $x_1, x_2, \ldots\ldots, x_n$ (the space parameters), and one dependent variable $u = u(t, x_1, x_2, \ldots\ldots, x_n)$, subject to:

$$\frac{\partial u}{\partial t} = k \sum_{i=1}^{n} \frac{\partial^2 u}{(\partial x_i)^2}$$

General Description

Suppose one has a function *u* that describes the temperature at a given location (*x*, *y*, *z*). This function will change over time as heat spreads throughout space. The heat equation is used to determine the change in the function *u* over time. The rate of change of ***u*** is proportional to the "curvature" of ***u***. Thus, the sharper the corner, the faster it is rounded off. Over time, the tendency is for peaks to be eroded, and valleys filled in. If *u* is linear in space (or has a constant gradient) at a given point, then *u* has reached steady-state and is unchanging at this point (assuming a constant thermal conductivity).

The image to the right is animated and describes the way heat changes in time along a metal bar. One of the interesting properties of the heat equation is the maximum principle that says that the maximum value of *u* is either earlier in time than the region of concern or on the edge of the region of concern. This is essentially saying that temperature comes either from some source or from earlier in time because heat permeates but is not created from nothingness. This is a property of parabolic partial differential equations and is not difficult to prove mathematically.

Another interesting property is that even if *u* has a discontinuity at an initial time $t = t_0$, the temperature becomes smooth as soon as $t > t_0$. For example, if a bar of metal has temperature 0 and another has temperature 100 and they are stuck together end to end, then very quickly the temperature at the point of connection will become 50 and the graph of the temperature will run smoothly from 0 to 50.

The heat equation is used in probability and describes random walks. It is also applied in financial mathematics for this reason.

It is also important in Riemannian geometry and thus topology: it was adapted by Richard S. Hamilton when he defined the Ricci flow that was later used by Grigori Perelman to solve the topological Poincaré conjecture.

The Physical Problem and the Equation

Derivation in One Dimension

The heat equation is derived from Fourier's law and conservation of energy (Cannon 1984). By Fourier's law, the rate of flow of heat energy per unit area through a surface is proportional to the negative temperature gradient across the surface,

$$\mathbf{q} = -k\nabla u$$

where k is the thermal conductivity and u is the temperature. In one dimension, the gradient is an ordinary spatial derivative, and so Fourier's law is

$$q = -k\frac{\partial u}{\partial x}$$

In the absence of work done, a change in internal energy per unit volume in the material, ΔQ, is proportional to the change in temperature, Δu (Δ is the ordinary difference operator with respect to time, not the Laplacian with respect to space). That is,

$$\Delta Q = c_p \rho \Delta u$$

where c_p is the specific heat capacity and ρ is the mass density of the material. Choosing zero energy at absolute zero temperature, this can be rewritten as

$$Q = c_p \rho u.$$

The increase in internal energy in a small spatial region of the material

$$x - \Delta x \leq \xi \leq x + \Delta x$$

over the time period

$$t - \Delta t \leq \tau \leq t + \Delta t$$

is given by

$$c_p\rho\int_{x-\Delta x}^{x+\Delta x}\left[u(\xi,t+\Delta t)-u(\xi,t-\Delta t)\right]d\xi = c_p\rho\int_{t-\Delta t}^{t+\Delta t}\int_{x-\Delta x}^{x+\Delta x}\frac{\partial u}{\partial \tau}d\xi d\tau$$

where the fundamental theorem of calculus was used. If no work is done and there are neither heat sources nor sinks, the change in internal energy in the interval $[x-\Delta x, x+\Delta x]$ is accounted for entirely by the flux of heat across the boundaries. By Fourier's law, this is

$$k\int_{t-\Delta t}^{t+\Delta t}\left[\frac{\partial u}{\partial x}(x+\Delta x,\tau)-\frac{\partial u}{\partial x}(x-\Delta x,\tau)\right]d\tau = k\int_{t-\Delta t}^{t+\Delta t}\int_{x-\Delta x}^{x+\Delta x}\frac{\partial^2 u}{\partial \xi^2}d\xi d\tau$$

again by the fundamental theorem of calculus. By conservation of energy,

$$\int_{t-\Delta t}^{t+\Delta t}\int_{x-\Delta x}^{x+\Delta x}\left[c_p\rho u_\tau - ku_{\xi\xi}\right]d\xi d\tau = 0.$$

This is true for any rectangle $[t-\Delta t, t+\Delta t] \times [x-\Delta x, x+\Delta x]$. By the fundamental lemma of the calculus of variations, the integrand must vanish identically:

$$c_p\rho u_t - ku_{xx} = 0.$$

Which can be rewritten as:

$$u_t = \frac{k}{c_p\rho}u_{xx},$$

or:

$$\frac{\partial u}{\partial t} = \frac{k}{c_p\rho}\left(\frac{\partial^2 u}{\partial x^2}\right)$$

which is the heat equation, where the coefficient (often denoted α)

$$\alpha = \frac{k}{c_p\rho}$$

is called the thermal diffusivity.

An additional term may be introduced into the equation to account for radiative loss of heat, which depends upon the excess temperature $u = T - T_s$ at a given point compared with the surroundings. At low excess temperatures, the radiative loss is approximately μu, giving a one-dimensional heat-transfer equation of the form

$$\frac{\partial u}{\partial t} = \frac{k}{c_p\rho}\left(\frac{\partial^2 u}{\partial x^2}\right) - \mu u.$$

At high excess temperatures, however, the Stefan–Boltzmann law gives a net radiative heat-loss proportional to $T^4 - T_s^4$, and the above equation is inaccurate. For large excess temperatures, $T^4 - T_s^4 \approx u^4$, giving a high-temperature heat-transfer equation of the form

$$\frac{\partial u}{\partial t} = \alpha\left(\frac{\partial^2 u}{\partial x^2}\right) - mu^4$$

where $m = \epsilon\sigma p / \rho A c_p$. Here, σ is Stefan's constant, ε is a characteristic constant of the material, p is the sectional perimeter of the bar and A is its cross-sectional area. However, using T instead of u gives a better approximation in this case.

Three-Dimensional Problem

In the special cases of wave propagation of heat in an isotropic and homogeneous medium in a 3-dimensional space, this equation is

$$\frac{\partial u}{\partial t} = \alpha \nabla^2 u = \alpha \left(\frac{\partial^2 u}{\partial x^2} + \frac{\partial^2 u}{\partial y^2} + \frac{\partial^2 u}{\partial z^2} \right) = \alpha (u_{xx} + u_{yy} + u_{zz})$$

where:

- $u = u(x, y, z, t)$ is temperature as a function of space and time;
- $\frac{\partial u}{\partial t}$ is the rate of change of temperature at a point over time;
- u_{xx}, u_{yy}, and u_{zz} are the second spatial derivatives (*thermal conductions*) of temperature in the x, y, and z directions, respectively;
- $\alpha = \frac{k}{c_p \rho}$ is the thermal diffusivity, a material-specific quantity depending on the *thermal conductivity* k, the *mass density* ρ, and the *specific heat capacity* c_p.

The heat equation is a consequence of Fourier's law of conduction (see heat conduction).

If the medium is not the whole space, in order to solve the heat equation uniquely we also need to specify boundary conditions for u. To determine uniqueness of solutions in the whole space it is necessary to assume an exponential bound on the growth of solutions.

Solutions of the heat equation are characterized by a gradual smoothing of the initial temperature distribution by the flow of heat from warmer to colder areas of an object. Generally, many different states and starting conditions will tend toward the same stable equilibrium. As a consequence, to reverse the solution and conclude something about earlier times or initial conditions from the present heat distribution is very inaccurate except over the shortest of time periods.

The heat equation is the prototypical example of a parabolic partial differential equation.

Using the Laplace operator, the heat equation can be simplified, and generalized to similar equations over spaces of arbitrary number of dimensions, as

$$u_t = \alpha \nabla^2 u = \alpha \Delta u,$$

where the Laplace operator, Δ or ∇^2, the divergence of the gradient, is taken in the spatial variables.

The heat equation governs heat diffusion, as well as other diffusive processes, such as particle diffusion or the propagation of action potential in nerve cells. Although they are not diffusive in nature, some quantum mechanics problems are also governed by a mathematical analog of the heat equation. It also can be used to model some phenomena arising in finance, like the Black–Scholes or the Ornstein-Uhlenbeck processes. The equation, and various non-linear analogues, has also been used in image analysis.

The heat equation is, technically, in violation of special relativity, because its solutions involve instantaneous propagation of a disturbance. The part of the disturbance outside the forward light cone can usually be safely neglected, but if it is necessary to develop a reasonable speed for the transmission of heat, a hyperbolic problem should be considered instead – like a partial differential equation involving a second-order time derivative. Some models of nonlinear heat conduction (which are also parabolic equations) have solutions with finite heat transmission speed.

Internal Heat Generation

The function u above represents temperature of a body. Alternatively, it is sometimes convenient to change units and represent u as the heat density of a medium. Since heat density is proportional to temperature in a homogeneous medium, the heat equation is still obeyed in the new units.

Suppose that a body obeys the heat equation and, in addition, generates its own heat per unit volume (e.g., in watts/litre - W/L) at a rate given by a known function q varying in space and time. Then the heat per unit volume u satisfies an equation

$$\frac{\partial u}{\partial t} = \alpha \left(\frac{\partial^2 u}{\partial x^2} + \frac{\partial^2 u}{\partial y^2} + \frac{\partial^2 u}{\partial z^2} \right) + \frac{1}{c_p \rho} q.$$

For example, a tungsten light bulb filament generates heat, so it would have a positive nonzero value for q when turned on. While the light is turned off, the value of q for the tungsten filament would be zero.

Solving the Heat Equation using Fourier Series

The following solution technique for the heat equation was proposed by Joseph Fourier in his treatise *Théorie analytique de la chaleur*, published in 1822. Let us consider the heat equation for one space variable. This could be used to model heat conduction in a rod. The equation is

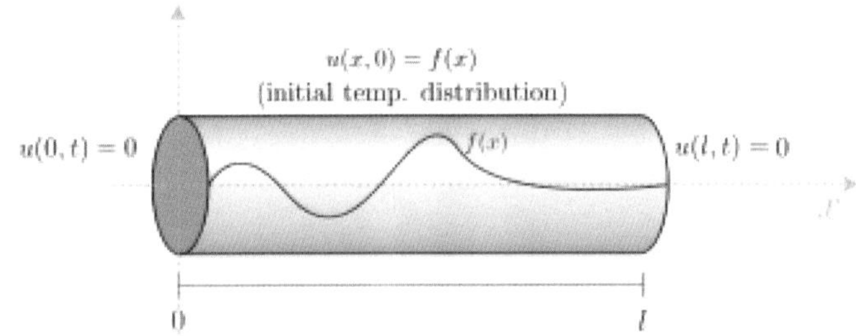

Idealized physical setting for heat conduction in a rod with homogeneous boundary conditions

$$u_t = \alpha u_{xx} \quad (1)$$

where $u = u(x, t)$ is a function of two variables x and t. Here

- x is the space variable, so $x \in [0, L]$, where L is the length of the rod.
- t is the time variable, so $t \geq 0$.

We assume the initial condition

$$u(x,0) = f(x) \quad \forall x \in [0, L] \quad (2)$$

where the function f is given, and the boundary conditions

$$u(0,t) = 0 = u(L,t) \quad \forall t > 0 \quad (3)$$

Let us attempt to find a solution of (1) that is not identically zero satisfying the boundary conditions (3) but with the following property: u is a product in which the dependence of u on x, t is separated, that is:

$$u(x,t) = X(x)T(t). \quad (4)$$

This solution technique is called separation of variables. Substituting u back into equation (1),

$$\frac{T'(t)}{\alpha T(t)} = \frac{X''(x)}{X(x)}.$$

Since the right hand side depends only on x and the left hand side only on t, both sides are equal to some constant value $-\lambda$. Thus:

$$T'(t) = -\lambda \alpha T(t) \qquad (5)$$

and

$$X''(x) = -\lambda X(x). \quad (6)$$

We will now show that nontrivial solutions for (6) for values of $\lambda \leq 0$ cannot occur:

1. Suppose that $\lambda < 0$. Then there exist real numbers B, C such that

   $$X(x) = Be^{\sqrt{-\lambda}\,x} + Ce^{-\sqrt{-\lambda}\,x}.$$

   From (3) we get $X(0) = 0 = X(L)$ and therefore $B = 0 = C$ which implies u is identically 0.

2. Suppose that $\lambda = 0$. Then there exist real numbers B, C such that $X(x) = Bx + C$. From equation (3) we conclude in the same manner as in 1 that u is identically 0.

3. Therefore, it must be the case that $\lambda > 0$. Then there exist real numbers A, B, C such that

   $$T(t) = Ae^{-\lambda \alpha t}$$

   and

   $$X(x) = B\sin(\sqrt{\lambda}\,x) + C\cos(\sqrt{\lambda}\,x).$$

   From (3) we get $C = 0$ and that for some positive integer n,

   $$\sqrt{\lambda} = n\frac{\pi}{L}.$$

This solves the heat equation in the special case that the dependence of u has the special form (4).

In general, the sum of solutions to (1) that satisfy the boundary conditions (3) also satisfies (1) and (3). We can show that the solution to (1), (2) and (3) is given by

$$u(x,t) = \sum_{n=1}^{\infty} D_n \sin\left(\frac{n\pi x}{L}\right) e^{-\frac{n^2\pi^2 \alpha t}{L^2}}$$

where

$$D_n = \frac{2}{L}\int_0^L f(x)\sin\left(\frac{n\pi x}{L}\right)dx.$$

Generalizing the Solution Technique

The solution technique used above can be greatly extended to many other types of equations. The idea is that the operator u_{xx} with the zero boundary conditions can be represented in terms of its eigenvectors. This leads naturally to one of the basic ideas of the spectral theory of linear self-adjoint operators.

Consider the linear operator $\Delta u = u_{xx}$. The infinite sequence of functions

$$e_n(x) = \sqrt{\frac{2}{L}} \sin\left(\frac{n\pi x}{L}\right)$$

for $n \geq 1$ are eigenvectors of Δ. Indeed,

$$\Delta e_n = -\frac{n^2\pi^2}{L^2} e_n.$$

Moreover, any eigenvector f of Δ with the boundary conditions $f(0) = f(L) = 0$ is of the form e_n for some $n \geq 1$. The functions e_n for $n \geq 1$ form an orthonormal sequence with respect to a certain inner product on the space of real-valued functions on $[0, L]$. This means

$$\langle e_n, e_m \rangle = \int_0^L e_n(x) e_m^*(x)dx = \delta_{mn}$$

Finally, the sequence $\{e_n\}_{n \in N}$ spans a dense linear subspace of $L^2((0, L))$. This shows that in effect we have diagonalized the operator Δ.

Heat Conduction in Non-homogeneous Anisotropic Media

In general, the study of heat conduction is based on several principles. Heat flow is a form of energy flow, and as such it is meaningful to speak of the time rate of flow of heat into a region of space.

- The time rate of heat flow into a region V is given by a time-dependent quantity $q_t(V)$. We assume q has a density Q, so that

  $$q_t(V) = \int_V Q(x,t)dx$$

- Heat flow is a time-dependent vector function H(x) characterized as follows: the time rate of heat flowing through an infinitesimal surface element with area dS and with unit normal vector n is

  $$\mathbf{H}(x) \cdot \mathbf{n}(x)dS$$

  Thus the rate of heat flow into V is also given by the surface integral

  $$q_t(V) = -\int_{\partial V} \mathbf{H}(x) \cdot \mathbf{n}(x)dS$$

  where n(x) is the outward pointing normal vector at x.

- The Fourier law states that heat energy flow has the following linear dependence on the temperature gradient

  $$\mathbf{H}(x) = -\mathbf{A}(x) \cdot \nabla u(x)$$

where A(x) is a 3 × 3 real matrix that is symmetric and positive definite.

- By the divergence theorem, the previous surface integral for heat flow into V can be transformed into the volume integral

$$q_t(V) = -\int_{\partial V} \mathbf{H}(x)\cdot\mathbf{n}(x)dS$$
$$= \int_{\partial V} \mathbf{A}(x)\cdot\nabla u(x)\cdot\mathbf{n}(x)dS$$
$$= \int_V \sum_{i,j} \partial_{x_i}\left(a_{ij}(x)\partial_{x_j}u(x,t)\right)dx$$

- The time rate of temperature change at x is proportional to the heat flowing into an infinitesimal volume element, where the constant of proportionality is dependent on a constant κ

$$\partial_t u(x,t) = \kappa(x)Q(x,t)$$

Putting these equations together gives the general equation of heat flow:

$$\partial_t u(x,t) = \kappa(x)\sum_{i,j}\partial_{x_i}\left(a_{ij}(x)\partial_{x_j}u(x,t)\right)$$

Remarks.

- The coefficient $\kappa(x)$ is the inverse of specific heat of the substance at x × density of the substance at x: $\kappa = 1/(\rho c_p)$.
- In the case of an isotropic medium, the matrix A is a scalar matrix equal to thermal conductivity k.
- In the anisotropic case where the coefficient matrix A is not scalar and/or if it depends on x, then an explicit formula for the solution of the heat equation can seldom be written down. Though, it is usually possible to consider the associated abstract Cauchy problem and show that it is a well-posed problem and/or to show some qualitative properties (like preservation of positive initial data, infinite speed of propagation, convergence toward an equilibrium, smoothing properties). This is usually done by one-parameter semigroups theory: for instance, if A is a symmetric matrix, then the elliptic operator defined by

$$Au(x) := \sum_{i,j}\partial_{x_i}a_{ij}(x)\partial_{x_j}u(x)$$

is self-adjoint and dissipative, thus by the spectral theorem it generates a one-parameter semigroup.

Fundamental Solutions

A fundamental solution, also called a *heat kernel*, is a solution of the heat equation corresponding to the initial condition of an initial point source of heat at a known position. These can be used to find a general solution of the heat equation over certain domains; see, for instance, (Evans 1998) for an introductory treatment.

In one variable, the Green's function is a solution of the initial value problem

$$\begin{cases} u_t(x,t) - ku_{xx}(x,t) = 0 & (x,t) \in \mathbf{R} \times (0,\infty) \\ u(x,0) = \delta(x) \end{cases}$$

where δ is the Dirac delta function. The solution to this problem is the fundamental solution

$$\Phi(x,t) = \frac{1}{\sqrt{4\pi kt}} \exp\left(-\frac{x^2}{4kt}\right).$$

One can obtain the general solution of the one variable heat equation with initial condition $u(x, 0) = g(x)$ for $-\infty < x < \infty$ and $0 < t < \infty$ by applying a convolution:

$$u(x,t) = \int \Phi(x-y,t)g(y)dy.$$

In several spatial variables, the fundamental solution solves the analogous problem

$$\begin{cases} u_t(\mathbf{x},t) - k\sum_{i=1}^{n} u_{x_i x_i}(\mathbf{x},t) = 0 & (\mathbf{x},t) \in \mathbf{R}^n \times (0,\infty) \\ u(\mathbf{x},0) = \delta(\mathbf{x}) \end{cases}$$

The n-variable fundamental solution is the product of the fundamental solutions in each variable; i.e.,

$$\Phi(\mathbf{x},t) = \Phi(x_1,t)\Phi(x_2,t)\ldots\Phi(x_n,t) = \frac{1}{\sqrt{(4\pi kt)^n}} \exp\left(-\frac{\mathbf{x}\cdot\mathbf{x}}{4kt}\right).$$

The general solution of the heat equation on R^n is then obtained by a convolution, so that to solve the initial value problem with $u(x, 0) = g(x)$, one has

$$u(\mathbf{x},t) = \int_{\mathbf{R}^n} \Phi(\mathbf{x}-\mathbf{y},t)g(\mathbf{y})d\mathbf{y}.$$

The general problem on a domain Ω in R^n is

$$\begin{cases} u_t(\mathbf{x},t) - k\sum_{i=1}^{n} u_{x_i x_i}(\mathbf{x},t) = 0 & (\mathbf{x},t) \in \Omega \times (0,\infty) \\ u(\mathbf{x},0) = g(\mathbf{x}) & \mathbf{x} \in \Omega \end{cases}$$

with either Dirichlet or Neumann boundary data. A Green's function always exists, but unless the domain Ω can be readily decomposed into one-variable problems, it may not be possible to write it down explicitly. Other methods for obtaining Green's functions include the method of images, separation of variables, and Laplace transforms (Cole, 2011).

Some Green's Function Solutions in 1D

A variety of elementary Green's function solutions in one-dimension are recorded here; many others are available elsewhere. In some of these, the spatial domain is $(-\infty,\infty)$. In others, it is the semi-infinite interval $(0,\infty)$ with either Neumann or Dirichlet boundary conditions. One further variation is that some of these solve the inhomogeneous equation

$$u_t = ku_{xx} + f.$$

where f is some given function of x and t.

Homogeneous Heat Equation

Initial value problem on $(-\infty, \infty)$

$$\begin{cases} u_t = ku_{xx} & (x,t) \in \quad \times (0,\infty) \\ u(x,0) \quad g(x) & IC \end{cases}$$

$$u(x,t) = \frac{1}{\sqrt{4\pi kt}} \int_{-\infty}^{\infty} \exp\left(-\frac{(x-y)^2}{4kt} \right) g(y)\,dy$$

This solution is the convolution with respect to the variable x of the fundamental solution

$$\Phi(x,t) := \frac{1}{\sqrt{4\pi kt}} \exp\left(-\frac{x^2}{4kt} \right),$$

and the function $g(x)$. (The Green's function number of the fundamental solution is X00.)

Therefore, according to the general properties of the convolution with respect to differentiation, $u = g * \Phi$ is a solution of the same heat equation, for

$$\left(\partial_t - k\partial_x^2\right)(\Phi * g) = \left[\left(\partial_t - k\partial_x^2\right)\Phi\right] * g = 0.$$

Moreover,

$$\Phi(x,t) = \frac{1}{\sqrt{t}}\Phi\left(\frac{x}{\sqrt{t}}\right)$$

$$\int_{-\infty}^{\infty} \Phi(x,t)dx = 1,$$

so that, by general facts about approximation to the identity, $\Phi(\cdot, t) * g \to g$ as $t \to 0$ in various senses, according to the specific g. For instance, if g is assumed bounded and continuous on R then $\Phi(\cdot, t) * g$ converges uniformly to g as $t \to 0$, meaning that $u(x, t)$ is continuous on R × $[0, \infty)$ with $u(x, 0) = g(x)$.

Initial value problem on $(0,\infty)$ with homogeneous Dirichlet boundary conditions

$$\begin{cases} u_t = ku_{xx} & (x,t) \in [0,\infty) \times (0,\infty) \\ u(x,0) = g(x) & IC \\ u(0,t) = 0 & BC \end{cases}$$

$$u(x,t) = \frac{1}{\sqrt{4\pi kt}} \int_0^{\infty} \left[\exp\left(-\frac{(x-y)^2}{4kt}\right) + \exp\left(-\frac{(x+y)^2}{4kt}\right)\right] g(y)dy$$

This solution is obtained from the preceding formula as applied to the data $g(x)$ suitably extended to R, so as to be an odd function, that is, letting $g(-x) := -g(x)$ for all x. Correspondingly, the solution of the initial value problem on $(-\infty,\infty)$ is an odd function with respect to the variable x for all values of t, and in particular it satisfies the homogeneous Dirichlet boundary conditions $u(0, t) = 0$. The Green's function number of this solution is X10.

Initial value problem on $(0,\infty)$ with homogeneous Neumann boundary conditions

$$\begin{cases} u_t = ku_{xx} & (x,t) \in [0,\infty) \times (0,\infty) \\ u(x,0) = g(x) & IC \\ u_x(0,t) = 0 & BC \end{cases}$$

$$u(x,t) = \frac{1}{\sqrt{4\pi kt}} \int_0^{\infty} \left[\exp\left(-\frac{(x-y)^2}{4kt}\right) + \exp\left(-\frac{(x+y)^2}{4kt}\right)\right] g(y)dy$$

This solution is obtained from the first solution formula as applied to the data *g(x)* suitably extended to R so as to be an even function, that is, letting $g(-x) := g(x)$ for all x. Correspondingly, the solution of the initial value problem on R is an even function with respect to the variable x for all values of $t > 0$, and in particular, being smooth, it satisfies the homogeneous Neumann boundary conditions $u_x(0, t) = 0$. The Green's function number of this solution is X20.

Problem on $(0,\infty)$ with homogeneous initial conditions and non-homogeneous Dirichlet boundary conditions

$$\begin{cases} u_t = ku_{xx} & (x,t) \in [0,\infty)\times(0,\infty) \\ u(x,0) = 0 & IC \\ u(0,t) = h(t) & BC \end{cases}$$

$$u(x,t) = \int_0^t \frac{x}{\sqrt{4\pi k(t-s)^3}} \exp\left(-\frac{x^2}{4k(t-s)}\right) h(s)ds, \qquad \forall x > 0$$

This solution is the convolution with respect to the variable t of

$$\psi(x,t) := -2k\partial_x \Phi(x,t) = \frac{x}{\sqrt{4\pi k t^3}} \exp\left(-\frac{x^2}{4kt}\right)$$

and the function $h(t)$. Since $\Phi(x, t)$ is the fundamental solution of

$$\partial_t - k\partial_x^2,$$

the function $\psi(x, t)$ is also a solution of the same heat equation, and so is $u := \psi * h$, thanks to general properties of the convolution with respect to differentiation. Moreover,

$$\psi(x,t) = \frac{1}{x^2}\psi\left(1, \frac{t}{x^2}\right)$$

$$\int_0^\infty \psi(x,t)dt = 1,$$

so that, by general facts about approximation to the identity, $\psi(x, \cdot) * h \to h$ as $x \to 0$ in various senses, according to the specific h. For instance, if h is assumed continuous on R with support in $[0, \infty)$ then $\psi(x, \cdot) * h$ converges uniformly on compacta to h as $x \to 0$, meaning that $u(x, t)$ is continuous on $[0, \infty) \times [0, \infty)$ with $u(0, t) = h(t)$.

Inhomogeneous Heat Equation

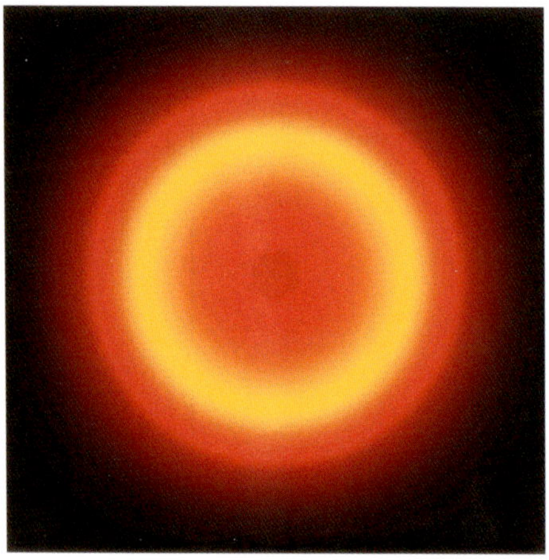

Depicted is a numerical solution of the nonhomogeneous heat equation. The equation has been solved with 0 initial and boundary conditions and a source term representing a stove top burner

Problem on (-∞,∞) homogeneous initial conditions

$$\begin{cases} u_t = ku_{xx} + f(x,t) & (x,t) \in \mathbf{R} \times (0,\infty) \\ u(x,0) = 0 & IC \end{cases}$$

$$u(x,t) = \int_0^t \int_{-\infty}^{\infty} \frac{1}{\sqrt{4\pi k(t-s)}} \exp\left(-\frac{(x-y)^2}{4k(t-s)} \right) f(y,s)\,dy\,ds$$

This solution is the convolution in R^2, that is with respect to both the variables x and t, of the fundamental solution

$$\Phi(x,t) := \frac{1}{\sqrt{4\pi kt}} \exp\left(-\frac{x^2}{4kt} \right)$$

and the function $f(x, t)$, both meant as defined on the whole R^2 and identically 0 for all $t \to 0$. One verifies that

$$\left(\partial_t - k\partial_x^2\right)(\Phi * f) = f,$$

which expressed in the language of distributions becomes

$$\left(\partial_t - k\partial_x^2\right)\Phi = \delta,$$

where the distribution δ is the Dirac's delta function, that is the evaluation at 0.

Problem on $(0,\infty)$ with homogeneous Dirichlet boundary conditions and initial conditions

$$\begin{cases} u_t = ku_{xx} + f(x,t) & (x,t) \in [0,\infty) \times (0,\infty) \\ u(x,0) = 0 & IC \\ u(0,t) = 0 & BC \end{cases}$$

$$u(x,t) = \int_0^t \int_0^\infty \frac{1}{\sqrt{4\pi k(t-s)}} \left(\exp\left(-\frac{(x-y)^2}{4k(t-s)} \right) - \exp\left(-\frac{(x+y)^2}{4k(t-s)} \right) \right) f(y,s)\,dy\,ds$$

This solution is obtained from the preceding formula as applied to the data $f(x, t)$ suitably extended to $\mathrm{R} \times [0,\infty)$, so as to be an odd function of the variable x, that is, letting $f(-x, t) := -f(x, t)$ for all x and t. Correspondingly, the solution of the inhomogeneous problem on $(-\infty,\infty)$ is an odd function with respect to the variable x for all values of t, and in particular it satisfies the homogeneous Dirichlet boundary conditions $u(0, t) = 0$.

Problem on $(0,\infty)$ with homogeneous Neumann boundary conditions and initial conditions

$$\begin{cases} u_t = ku_{xx} + f(x,t) & (x,t) \in [0,\infty) \times (0,\infty) \\ u(x,0) = 0 & IC \\ u_x(0,t) = 0 & BC \end{cases}$$

$$u(x,t) = \int_0^t \int_0^\infty \frac{1}{\sqrt{4\pi k(t-s)}} \left(\exp\left(-\frac{(x-y)^2}{4k(t-s)} \right) + \exp\left(-\frac{(x+y)^2}{4k(t-s)} \right) \right) f(y,s)\,dy\,ds$$

This solution is obtained from the first formula as applied to the data $f(x, t)$ suitably extended to $\mathrm{R} \times [0,\infty)$, so as to be an even function of the variable x, that is, letting $f(-x, t) := f(x, t)$ for all x and t. Correspondingly, the solution of the inhomogeneous problem on $(-\infty,\infty)$ is an even function with respect to the variable x for all values of t, and in particular, being a smooth function, it satisfies the homogeneous Neumann boundary conditions $u_x(0, t) = 0$.

Examples

Since the heat equation is linear, solutions of other combinations of boundary conditions, inhomogeneous term, and initial conditions can be found by taking an appropriate linear combination of the above Green's function solutions.

For example, to solve

$$\begin{cases} u_t = ku_{xx} + f & (x,t) \in \mathbf{R} \times (0,\infty) \\ u(x,0) = g(x) & IC \end{cases}$$

let $u = w + v$ where w and v solve the problems

$$\begin{cases} v_t = kv_{xx} + f, w_t = kw_{xx} & (x,t) \in \mathbf{R} \times (0,\infty) \\ v(x,0) = 0, w(x,0) = g(x) & IC \end{cases}$$

Similarly, to solve

$$\begin{cases} u_t = ku_{xx} + f & (x,t) \in [0,\infty) \times (0,\infty) \\ u(x,0) = g(x) & IC \\ u(0,t) = h(t) & BC \end{cases}$$

let $u = w + v + r$ where w, v, and r solve the problems

$$\begin{cases} v_t = kv_{xx} + f, w_t = kw_{xx}, r_t = kr_{xx} & (x,t) \in [0,\infty) \times (0,\infty) \\ v(x,0) = 0,\ w(x,0) = g(x),\ r(x,0) = 0 & IC \\ v(0,t) = 0,\ w(0,t) = 0,\ r(0,t) = h(t) & BC \end{cases}$$

Mean-value Property for the Heat Equation

Solutions of the heat equations

$$(\partial_t - \Delta)u = 0$$

satisfy a mean-value property analogous to the mean-value properties of harmonic functions, solutions of

$$\Delta u = 0\ ,$$

though a bit more complicated. Precisely, if u solves

$$(\partial_t - \Delta)u = 0$$

and

$$(x,t) + E_\lambda \subset dom(u)$$

then

$$u(x,t) = \frac{\lambda}{4}\int_{E_\lambda} u(x-y, t-s)\frac{|y|^2}{s^2}\, ds\, dy,$$

where E_λ is a "heat-ball", that is a super-level set of the fundamental solution of the heat equation:

$$E_\lambda := \{(y,s):\Phi(y,s) > \lambda\},$$

$$\Phi(x,t) := (4t\pi)^{-\frac{n}{2}} \exp\left(-\frac{|x|^2}{4t}\right).$$

Notice that

$$\text{diam}(E_1) = o(1)$$

as $\lambda \to \infty$ so the above formula holds for any (x, t) in the (open) set dom(u) for λ large enough. This can be shown by an argument similar to the analogous one for harmonic functions.

Steady-state Heat Equation

The steady-state heat equation is by definition not dependent on time. In other words, it is assumed conditions exist such that:

$$\frac{\partial u}{\partial t} = 0$$

This condition depends on the time constant and the amount of time passed since boundary conditions have been imposed. Thus, the condition is fulfilled in situations in which the *time equilibrium constant is fast enough* that the more complex time-dependent heat equation can be approximated by the steady-state case. Equivalently, the steady-state condition exists for all cases in which *enough time has passed* that the thermal field u no longer evolves in time.

In the steady-state case, a spatial thermal gradient may (or may not) exist, but if it does, it does not change in time. This equation therefore describes the end result in all thermal problems in which a source is switched on (for example, an engine started in an automobile), and enough time has passed for all permanent temperature gradients to establish themselves in space, after which these spatial gradients no longer change in time (as again, with an automobile in which the engine has been running for long enough). The other (trivial) solution is for all spatial temperature gradients to disappear as well, in which case the temperature become uniform in space, as well.

The equation is much simpler and can help to understand better the physics of the materials without focusing on the dynamic of the heat transport process. It is widely used for simple engineering problems assuming there is equilibrium of the temperature fields and heat transport, with time.

Steady-state condition:

$$\frac{\partial u}{\partial t} = 0$$

The steady-state heat equation for a volume that contains a heat source (the inhomogeneous case), is the Poisson's equation:

$$-k\nabla^2 u = q$$

where u is the temperature, k is the thermal conductivity and q the heat-flux density of the source.

In electrostatics, this is equivalent to the case where the space under consideration contains an electrical charge.

The steady-state heat equation without a heat source within the volume (the homogeneous case) is the equation in electrostatics for a volume of free space that does not contain a charge. It is described by Laplace's equation:

$$\nabla^2 u = 0$$

Applications

Particle Diffusion

One can model particle diffusion by an equation involving either:

- the volumetric concentration of particles, denoted *c*, in the case of collective diffusion of a large number of particles, or
- the probability density function associated with the position of a single particle, denoted *P*.

In either case, one uses the heat equation

$$c_t = D\Delta c,$$

or

$$P_t = D\Delta P.$$

Both *c* and *P* are functions of position and time. *D* is the diffusion coefficient that controls the speed of the diffusive process, and is typically expressed in meters squared over second. If the diffusion coefficient *D* is not constant, but depends on the concentration *c* (or *P* in the second case), then one gets the nonlinear diffusion equation.

Brownian Motion

Let the stochastic process X be the solution of the stochastic differential equation

$$\begin{cases} dX_t = \sqrt{2k}\, dB_t \\ X_0 = 0 \end{cases}$$

where B is the Wiener process (standard Brownian motion). Then the probability density function of X is given at any time t by

$$\frac{1}{\sqrt{4\pi kt}}\exp\left(-\frac{x^2}{4kt}\right)$$

which is the solution of the initial value problem

$$\begin{cases} u_t(x,t) - ku_{xx}(x,t) = 0, & (x,t) \in \mathbb{R}\times(0,+\infty) \\ u(x,0) = \delta(x) \end{cases}$$

where δ is the Dirac delta function.

Schrödinger Equation for a Free Particle

With a simple division, the Schrödinger equation for a single particle of mass m in the absence of any applied force field can be rewritten in the following way:

$$\psi_t = \frac{i\hbar}{2m}\Delta\psi,$$

where i is the imaginary unit, $\hbar$ is the reduced Planck's constant, and ψ is the wave function of the particle.

This equation is formally similar to the particle diffusion equation, which one obtains through the following transformation:

$$c(\mathbf{R},t) \to G(\mathbf{R},t)$$
$$D \to \frac{i\hbar}{2m}$$

Applying this transformation to the expressions of the Green functions determined in the case of particle diffusion yields the Green functions of the Schrödinger equation, which in turn can be used to obtain the wave function at any time through an integral on the wave function at $t = 0$:

$$\psi(\mathbf{R},t) = \int \psi(\mathbf{R}^0, t=0)G(\mathbf{R}-\mathbf{R}^0,t)dR_x^0\,dR_y^0\,dR_z^0,$$

with

$$G(\mathbf{R},t) = \left(\frac{m}{2\pi i\hbar t}\right)^{3/2} e^{-\frac{\mathbf{R}^2 m}{2i\hbar t}}.$$

Remark: this analogy between quantum mechanics and diffusion is a purely formal one. Physically, the evolution of the wave function satisfying Schrödinger's equation might have an origin other than diffusion.

Thermal Diffusivity in Polymers

A direct practical application of the heat equation, in conjunction with Fourier theory, in spherical coordinates, is the prediction of thermal transfer profiles and the measurement of the thermal diffusivity in polymers (Unsworth and Duarte). This dual theoretical-experimental method is applicable to rubber, various other polymeric materials of practical interest, and microfluids. These authors derived an expression for the temperature at the center of a sphere T_C

$$\frac{T_C - T_S}{T_0 - T_S} = 2\sum_{n=1}^{\infty}(-1)^{n+1}\exp\left(-\frac{n^2\pi^2\alpha t}{L^2}\right)$$

where T_0 is the initial temperature of the sphere and T_S the temperature at the surface of the sphere, of radius L. This equation has also found applications in protein energy transfer and thermal modeling in biophysics.

Further Applications

The heat equation arises in the modeling of a number of phenomena and is often used in financial mathematics in the modeling of options. The famous Black–Scholes option pricing model's differential equation can be transformed into the heat equation allowing relatively easy solutions from a familiar body of mathematics. Many of the extensions to the simple option models do not have closed form solutions and thus must be solved numerically to obtain a modeled option price. The equation describing pressure diffusion in a porous medium is identical in form with the heat equation. Diffusion problems dealing with Dirichlet, Neumann and Robin boundary conditions have closed form analytic solutions (Thambynayagam 2011). The heat equation is also widely used in image analysis (Perona & Malik 1990) and in machine-learning as the driving theory behind scale-space or graph Laplacian methods. The heat equation can be efficiently solved numerically using the implicit Crank–Nicolson method of (Crank & Nicolson 1947). This method can be extended to many of the models with no closed form solution.

An abstract form of heat equation on manifolds provides a major approach to the Atiyah–Singer index theorem, and has led to much further work on heat equations in Riemannian geometry.

Heat Kernel

There are many ways to introduce the heat kernel and each has its own advantages. Let us mention three distinct viewpoints.

The first is to view the heat kernel as the fundamental solution of a parabolic partial differential equation, the heat diffusion equation

$$(\partial_t - \Delta)u = 0.$$

That is, we seek a smooth function $(t,x,y) \mapsto p(t,x,y)$ defined on $(0,\infty) \times M \times M$ such that, for each $y \in M, p(t,x,y)$ is a solution of above equation and for any $\phi \in C_c^\infty(M)$ (smooth compactly supported function on M), $u(t,x) = \int_M p(t,x,y)\phi(y)dy$ tends to $\phi(x)$ as t tends to 0. In other words, the heat kernel allows us to solve the Cauchy initial value problem for above equation.

A second possibility is to consider Δ (defined, say, on $C_c^\infty(M)$) as a symmetric operator on $L^2(M, \mu)$ whose Friederichs extension generates a strongly continuous one parameter contraction semigroup H_t on $L^2(M, \mu)$ (the heat diffusion semigroup). This semigroup admits a transition function (a priori, this is a measure in the second variable) *p(t, x, dy)* such that $H_t f(x) = \int_M f(y)p(t,x,dy)$. The heat kernel (if it exists) is then obtained as the density of p(t, x, dy) with respect to the underlying measure $\mu(dy) = dy$.

A third approach is to view *p(t, x, dy)* as the distribution at time t of a stochastic process $(X_t)_{t>0}$ started at x (the Brownian motion driven by Δ on M). These different viewpoints are related by the formula

$$u(t,x) = H_t u_0(x) - \mathrm{E}_x(u_0(X_t))$$

where u is the solution of the Cauchy initial value problem with initial value condition u_0 (say, in $C_c^\infty(M)$).

Heat kernel in R^n

The classical Gauss-Weierstrass heat kernel is the following function

$$p_t(x,y) = \frac{1}{(4\pi t)^{n/2}} \exp\left(-\frac{|x-y|^2}{4t}\right),$$

where $x, y \in \mathbb{R}^n$ and t > 0. This function is a fundamental solution of the heat equation

$$\frac{\partial u}{\partial t} = \Delta u,$$

where $\Delta = \sum_{i=1}^n \frac{\partial^2}{\partial x_i^2}$ is the Laplace operator. Moreover, if f is a continuous bounded function on $\mathbb{R}^n$ then the function

$$u(t,x) = \int_{\mathbb{R}^n} p_t(x,y) f(y) dy$$

solves the Cauchy problem

$$\begin{cases} \dfrac{\partial u}{\partial t} = \Delta u \\ u \ \ (0,x) = f(x) \end{cases}.$$

This also can be written in the form

$U(t,.)=\exp(-t\mathscr{L})f,$

where $\mathscr{L}$ here is a self-adjoint extension of $-\Delta$ in $L^2(R^n)$ and $\exp(-t\mathscr{L})$ is understood in the sense of the functional calculus of self-adjoint operators. This means that $p_t(x, y)$ is the integral kernel of the operator $\exp(-t\mathscr{L})$.

The function $p_t(x, y)$ has also a probabilistic meaning: it is the transition density of Brownian motion $\{X_t\}_{t>0}$ in $\mathbb{R}^n$. The graph of $p_t(x, 0)$ as a function of x is shown here:

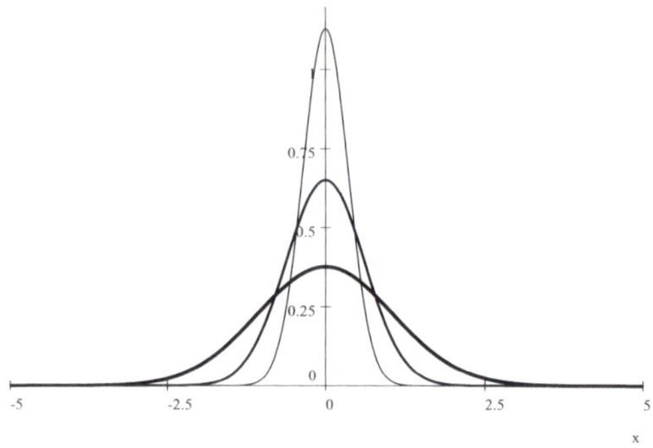

The Gauss-Weierstrass heat kernel at different values of t

The term $\frac{|x-y|^2}{t}$ determines the space/time scaling: if $|x-y|^2 \leq Ct$ then $p_t(x, y)$ is comparable with $p_t(x, x)$, that is, the probability density in the $C\sqrt{t}-$ neighborhood of x is nearly constant.

Heat Kernels on Riemannian Manifolds

Let (M, g) be a connected Riemannian manifold, and Δ be the Laplace-Beltrami operator on *M*. Then the heat kernel $p_t(x, y)$ is can be defined as the integral kernel of the heat semigroup $\{\exp(-t\mathscr{L})\}t>0$ where $\mathscr{L}$ is the Dirichlet Laplace operator, that is, the minimal self-adjoint extension of $-\Delta$ in $L^2(M, \mu)$, and μ is the Riemannian volume. Alternatively, $p_t(x, y)$ is the minimal positive fundamental solution of the heat equation

$$\frac{\partial u}{\partial t} = u.$$

The function $p_t(x, y)$ can be used to define Brownian motion $\{X_t\}_{t>0}$ on M. Namely,

$\{X_t\}_{t\geq 0}$ is a diffusion process (that is, a Markov process with continuous trajectories), such that

$$P_x(X_t \in A) = \int_A p_t(x, y)d\mu(y)$$

for any Borel set $A \subset M$.

Let $d(x, y)$ be the geodesic distance on M. It turns out that the Gaussian type space/time scaling $\frac{d^2(x,y)}{t}$ appears in heat kernel estimates on general Riemannian manifolds:

1. (Varadhan) For an arbitrary Riemannian manifold,

$$\log p_t(x, y) \sim -\frac{d^2(x,y)}{4t} \quad as \quad t \to 0,$$

2. (*Davies*) For an arbitrary manifold M, for any two measurable sets $A, B \subset M$

$$\int_A \int_B p_t(x, y)d\mu(x)d\mu(y) \leq \sqrt{\mu(A)\mu(B)} \exp\left(-\frac{d^2(A,B)}{4t}\right)$$

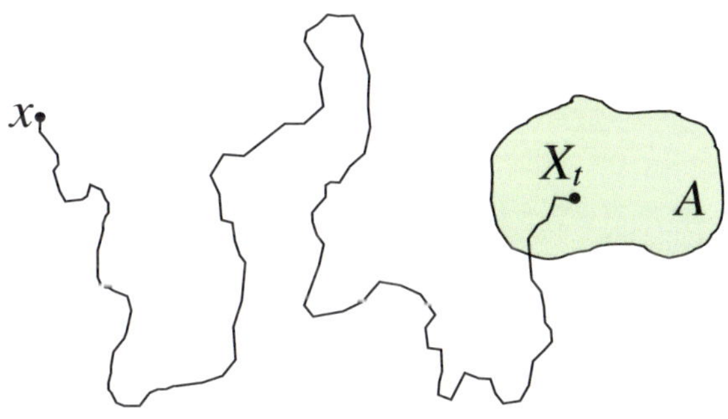

The Brownian motion X_t hits a set A

Technically, all these results depend upon the following property of the geodesic distance: $|\nabla d| \leq 1$.

Fourier's Law for Heat Conduction

An empirical relationship between the conduction rate in a material and the temperature gradient in the direction of energy flow, first formulated by Fourier in 1822 who concluded that "the heat flux resulting from thermal conduction is proportional to the magnitude of the temperature gradient and opposite to it in sign". For a unidirectional conduction process this observation may be expressed as:

$$\dot{q}_x = -\lambda \frac{\mathrm{dT}}{\mathrm{dx}}$$

where the vector $\dot{q}_x$ is the heat flux (W/m^2) in the positive x-direction, dT/dx is the (negative) temperature gradient (K/m) in the direction of heat flow (i.e., conduction occurs in the direction of decreasing temperature and the minus sign confirms this thermodynamic axiom) and the proportionality constant λ is the Thermal Conductivity of the material (W/mK). Fourier's Law thus provides the definition of thermal conductivity and forms the basis of many methods of determining its value. Fourier's Law, as the basic rate equation of the conduction process, when combined with the principle of conservation of energy, also forms the basis for the analysis of most Conduction problems.

If the temperature decreases with x, q will be positive and will flow in the direction of x. If the temperature increases with x, q will be negative, and will flow opposite to the direction of x.

Integral Form

By integrating the differential form over the material's total surface S, we arrive at the integral form of Fourier's law:

$$\frac{\partial Q}{\partial t} = -k \oiint_S \nabla T \cdot dS$$

where (including the SI units):

- $\frac{\partial Q}{\partial t}$ is the amount of heat transferred per unit time (in W), and
- dS is an oriented surface area element (in m^2)

The above differential equation, when integrated for a homogeneous material of 1-D geometry between two endpoints at constant temperature, gives the heat flow rate as:

$$\frac{Q}{\Delta t} = -kA \frac{\Delta T}{\Delta x}$$

where

A is the cross-sectional surface area,

ΔT is the temperature difference between the ends,

Δx is the distance between the ends.

This law forms the basis for the derivation of the heat equation.

Conductance

Writing

$$U = \frac{k}{\Delta x},$$

where U is the conductance, in W/(m^2 K).

Fourier's law can also be stated as:

$$\frac{\Delta Q}{\Delta t} = UA(-\Delta T).$$

The reciprocal of conductance is resistance, R, given by:

$$R = \frac{1}{U} = \frac{\Delta x}{k} = \frac{A(-\Delta T)}{\frac{\Delta Q}{\Delta t}}.$$

Resistance is additive when several conducting layers lie between the hot and cool regions, because A and Q are the same for all layers. In a multilayer partition, the total conductance is related to the conductance of its layers by:

$$\frac{1}{U} = \frac{1}{U_1} + \frac{1}{U_2} + \frac{1}{U_3} + \cdots$$

So, when dealing with a multilayer partition, the following formula is usually used:

$$\frac{\Delta Q}{\Delta t} = \frac{A(-\Delta T)}{\frac{\Delta x_1}{k_1} + \frac{\Delta x_2}{k_2} + \frac{\Delta x_3}{k_3} + \cdots}.$$

For heat conduction from one fluid to another through a barrier, it is sometimes important to consider the conductance of the thin film of fluid that remains stationary next to the barrier. This thin film of fluid is difficult to quantify because its characteristics depend upon complex conditions of turbulence and viscosity—but when dealing with thin high-conductance barriers it can sometimes be quite significant.

Intensive-property Representation

The previous conductance equations, written in terms of extensive properties, can be reformulated in terms of intensive properties. Ideally, the formulae for conductance should produce a quantity with dimensions independent of distance, like Ohm's Law for electrical resistance, $R = V / I$, and conductance, $G = I / V$.

From the electrical formula: $R = \rho x / A$, where ρ is resistivity, x is length, and A is cross-sectional area, we have $G = kA / x$, where G is conductance, k is conductivity, x is length, and A is cross-sectional area.

For Heat,

$$U = \frac{kA}{\Delta x},$$

where U is the conductance.

Fourier's law can also be stated as:

$$\dot{Q} = U \Delta T,$$

analogous to Ohm's law, $I = V / R$ or $I = VG$.

The reciprocal of conductance is resistance, R, given by:

$$R = \frac{\Delta T}{\dot{Q}},$$

analogous to Ohm's law, $R = V / I$.

The rules for combining resistances and conductances (in series and in parallel) are the same for both heat flow and electric current.

Cylindrical Shells

Conduction through cylindrical shells (e.g. pipes) can be calculated from the internal radius, r_1, the external radius, r_2, the length, ℓ, and the temperature difference between the inner and outer wall, $T_2 - T_1$.

The surface area of the cylinder is $A_r = 2\pi r \ell$

When Fourier's equation is applied:

$$\dot{Q} = -kA_r \frac{dT}{dr} = -2k\pi r \ell \frac{dT}{dr}$$

and rearranged:

$$\dot{Q} \int_{r_1}^{r_2} \frac{1}{r} dr = -2k\pi\ell \int_{T_1}^{T_2} dT$$

then the rate of heat transfer is:

$$\dot{Q} = 2k\pi\ell \frac{T_1 - T_2}{\ln(r_2 / r_1)}$$

the thermal resistance is:

$$R_c = \frac{\Delta T}{\dot{Q}} = \frac{\ln(r_2 / r_1)}{2\pi k\ell}$$

and $\dot{Q} = 2\pi k\ell r_m \frac{T_1 - T_2}{r_2 - r_1}$, where $r_m = \frac{r_2 - r_1}{\ln(r_2 / r_1)}$. It is important to note that this is the log-mean radius.

Spherical

The conduction through a spherical shell with internal radius, r_1, and external radius, r_2, can be calculated in a similar manner as for a cylindrical shell.

The surface area of the sphere is: $A = 4\pi r^2$.

Solving in a similar manner as for a cylindrical shell produces:

$$\dot{Q} = 4k\pi \frac{T_1 - T_2}{1/r_1 - 1/r_2} = 4k\pi \frac{(T_1 - T_2)r_1 r_2}{r_2 - r_1}$$

Thermal Conductivity

In physics, thermal conductivity, k, is the property of a material's ability to conduct heat. It appears primarily in Fourier's Law for heat conduction. Thermal conductivity is measured in watts per kelvin-meter ($W \cdot K^{-1} \cdot m^{-1}$, i.e. W/(K·m) or in IP units ($Btu/(hr^{-1} \cdot ft^{-1} \cdot F^{-1})$), i.e. Btu/(hr·ft·F). Multiplied by a temperature difference (in kelvins, K) and an area (in square meters, m^2), and divided by a thickness (in meters, m), the thermal conductivity predicts the rate of energy loss (in watts, W) through a piece of material. In the window building industry "thermal conductivity" is expressed as the U-Factor, which measures the rate of heat transfer and tells you how well the window insulates. U-factor values are generally recorded in IP units (Btu/(hr·ft·F)) and usually range from 0.15 to 1.25. The lower the U-factor, the better the window insulates.

The Reciprocal of Thermal Conductivity is Thermal Resistivity.

Heat transfer occurs at a higher rate across materials of high thermal conductivity than those of low thermal conductivity. Materials of high thermal conductivity are widely used in heat sink applications and materials of low thermal conductivity are used as thermal insulation. Thermal conductivity of materials is temperature dependent. The reciprocal of thermal conductivity is called thermal resistivity. Metals with high thermal conductivity, e.g. copper, exhibit high electrical conductivity. The heat generated in high thermal conductivity materials is rapidly conducted away from the region of the weld. For metallic materials, the electrical and thermal conductivity correlate positively, i.e. materials with high electrical conductivity (low electrical resistance) exhibit high thermal conductivity.

Measurement

There are a number of ways to measure thermal conductivity. Each of these is suitable for a limited range of materials, depending on the thermal properties and the medium temperature. There is a distinction between steady-state and transient techniques.

In general, steady-state techniques are useful when the temperature of the material does not change with time. This makes the signal analysis straightforward (steady state implies constant signals). The disadvantage is that a well-engineered experimental setup is usually needed. The Divided Bar (various types) is the most common device used for consolidated rock samples.

The transient techniques perform a measurement during the process of heating up. Their advantage is quicker measurements. Transient methods are usually carried out by needle probes.

Thermal Diffusivity

Thermal diffusivity is a combination of physical properties (conductivity, density and specific heat capacity) ($\lambda/\rho c_p$) denoted by κ, naturally arising in the derivation of the *conduction equation* and having physical significance in the context of transient conduction processes. It may be regarded as the ratio of the Thermal Conductivity of a material to the specific heat capacity of that material. Thus copper, with a high value of κ ($= 11 \times 10^{-5}$ $m^2 \cdot s^{-1}$), has a high conduction rate relative to its heat storage capacity and so responds quickly to changes in temperature. Conversely insulating materials may have κ values 100 times smaller, these low κ values being a measure of the high thermal inertia of such materials. Values of κ are often calculated from the widely available data on λ, ρ and c_p but directly determined values are also available.

For a homogenous material with constant properties, below given heat equation with three physical properties is expressed as second equation with just one coefficient.

$$pc_p \frac{\partial T}{\partial t} = k\left(\frac{\partial^2 T}{\partial x^2} + \frac{\partial^2 T}{\partial y^2} + \frac{\partial^2 T}{\partial z^2}\right)$$

$$\frac{\partial T}{\partial t} = \alpha\left(\frac{\partial^2 T}{\partial x^2} + \frac{\partial^2 T}{\partial y^2} + \frac{\partial^2 T}{\partial z^2}\right)$$

A material with a high thermal diffusivity (such as silver) is a good diffuser of thermal energy while a material with a low thermal diffusivity (such as plastic) is much slower at diffusing thermal energy. If the thermal environment around a material changes, heat must flow in or out of the material until thermal equilibrium is achieved, assuming the environment is constant after the change. Materials with a high thermal diffusivity will achieve thermal equilibrium faster than materials with low thermal diffusivity.

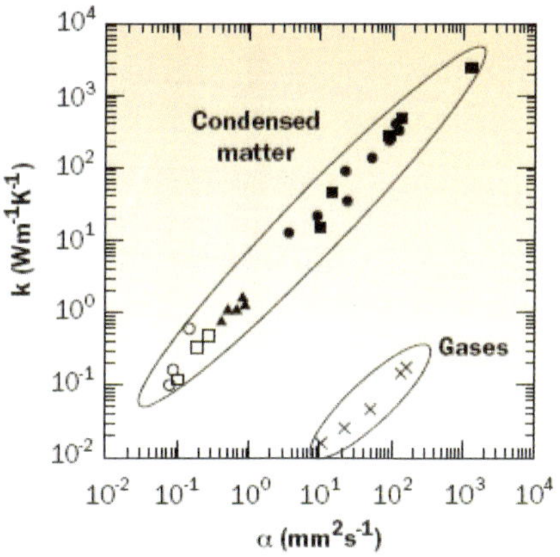

Thermal conductivity versus thermal diffusivity for a wide variety of homogeneous materials (closed circles, metals; squares, ceramics; triangles, glasses; open squares, polymers; open circles, liquids; and crosses, gases)

Figure above compares the thermal conductivity to thermal diffusivity for a wide range of materials. Notice that the data points noted as condensed matter can be described as being near a straight line. This is because the range of heat capacity per unit volume for condensed matter (liquids and solids) is small (the range of heat capacity per volume for condensed matter is from 1×10^{-6} J/m^3 -K to 4×10^{-6} J/m^3-K). With respect to thermal diffusivity, gases and condensed matter are different. For example, air has a low thermal conductivity but a relatively high thermal diffusivity this means that while air can only absorb a relatively small amount of thermal energy, it is effective at diffusing the energy.

Table: Thermal Diffusivity for Selected Materials

Material	Thermal Diffusivity (cm²/sec) @300 K
Silver	1.74
Gold	1.27
Copper	1.15
Aluminum	0.97
Silicon	0.88
Aluminum Alloy 6061-T6	0.64
Tin	0.40

Iron	0.23
Aluminum oxide	0.12
Stainless Steel 304A	0.042
Quartz	0.014
Silicon Dioxide (Polycrystalline)	0.0083
Water	0.0014
Polyvinyl Chloride (PVC)	0.0008
Alcohol	0.0007
Air	0.19

Thermal Effusivity

Thermal Effusivity, is a value to describe the behavior of the heat storing or dissipating capability of materials. It is the square root of the product of the material's thermal conductivity and its volumetric heat capacity:

$$\sqrt{Kpc_p}$$

K: Thermal Conductivity (W / m·K)

ρ: Density (kg/m²)

c_p: Specific Heat Capacity (J / kg·K)

Thermal effusivity is used to describe heat transition behavior of two objects, when the two objects are contacted to each other. When the two materials have the same thermal effusivity, the heat transition behavior will equal to a unique object as if there are no junction between the two objects.

A typical example is the human skin. When we touch different materials for a period, we will have different feelings. We feel cool with some materials, and feel warm with some others. Let's say, hand on a metal pipe in the park, we feel cool, and a cotton cloth is warm. To the physical measurement, the metal pipe has higher thermal effusivity and the cotton cloth is lower. Thermal effusivity is a unit which can express the real physical amount, instead of the feeling of 'cooler' or 'warmer'.

The traditional thermal conductivity analyzer needs a series of calculations to get the thermal effusivity, and even the thermal conductivity need to be calculated from the thermal resistance value, which the known density and specific heat capacity. Typically, these thermal conductivity analyzer takes a long time to get the final conductivity and effusivity.

Thermal effusivity (from low to high) for selected materials at room temperature. The penetration depth δ is tabulated at 1s after excitation.

	ρ	c_p	k	ρc_p	a	e	δ (1s)
	kg/m³	J/kgK	W/mK	J/m³K*10⁶	mm²/s	w/cm²/k/s^0.5	mm
air	1.3	1004	0.03	0.001	19.2	0.0006	8.8
wool	100	1500	0.04	0.15	0.23	0.007	1.0
balsa wood	130	2301	0.05	0.30	0.17	0.012	0.8
polyvinyl chloride	1500	1674	0.17	2.51	0.07	0.06	0.5
skin	1000	2500	0.40	2.50	0.16	0.10	0.8
quartz	2200	745	1.40	1.64	0.85	0.15	1.8
silicon oxide	2200	745	1.40	1.64	0.85	0.15	1.8
water	1000	4184	0.60	4.18	0.14	0.16	0.8
ice	917	4217	2.10	3.87	0.54	0.28	1.5
aluminum oxide	2200	778	18	1.71	10.5	0.56	6.5
stainless steel (CrNi)	8000	502	15	4.02	3.73	0.78	3.9
tin	7310	226	61	1.65	36.9	1.00	12.2
silicon	2330	703	126	1.64	76.9	1.44	17.5
iron	7870	448	72	3.52	20.4	1.59	9.0
aluminum alloy (7079)	2740	795	121	2.18	55.5	1.62	14.9
aluminum	2698	921	226	2.48	91	2.37	19.1
gold	19300	128	320	2.47	129	2.81	22.8
copper	8940	385	396	3.44	115	3.69	21.5

If two semi-infinite bodies initially at temperatures T_1 and T_2 are brought in perfect thermal contact, the temperature at the contact surface T_m will be given by their relative effusivities.

$$T_m = T_1 + (T_2 - T_1)\frac{e_2}{(e_2 + e_1)}$$

This expression is valid for all times for semi-infinite bodies in perfect thermal contact. It is also a good first guess for the initial contact temperature for finite bodies.

Direct measurement of thermal effusivity may be performed via specialty sensors, as pictured to the right.

Thermal Grease

Thermal grease, also called thermal paste or thermal compound, is a substance used to promote better heat conduction between two surfaces and is commonly used between a microprocessor and a heatsink. The top surface of most microprocessors is not perfectly flat. Some have microscopic grooves and others might even have a slight curve, which produces air gaps between the microprocessor and the heatsink and reduces the cooling performance of the heatsink. The air gaps are filled by applying a thin layer of thermal grease to the top of the microprocessor and the base of the heatsink.

There are two types of thermal grease: non-conductive and conductive. The non-conductive thermal grease types includes silicone and zinc thermal greases. The conductive types includes silver, copper, and aluminum-based greases. These are considered superior, but must be used carefully or they can cause electrical shorts if applied to the microprocessor's pin s or the electrical pathways on the circuit board.

Thermal Paste Application

Recommended Paste Thickness

Given the various parameters that influence thermal resistance, it should be understood that the target paste thickness will vary depending on module type. The manufacturer of the power module should qualify the module with an industry standard thermal paste and then specify a thickness. This paste has traditionally been Wacker P12 and the target thicknesses are given in the respective Technical Explanations/Mounting Instructions for the product line in question (e.g. MiniSKiiP, SEMITOP, etc.). These thicknesses are valid for any thermal paste with similar physical characteristics to Wacker P12 (i.e. silicone paste with oxide carriers and density of ~2.1g/cm^3) and are usually specified over a 5-20μm wide range.

It can be seen from the previous discussion that only enough paste should be applied to fill air-filled voids. Just adding a thin layer of paste will quickly reduce the thermal resistance in the system, but adding additional paste beyond this will slowly raise the thermal resistance since the thermal conductivity of the paste is lower than the two metals in the interface figure below.

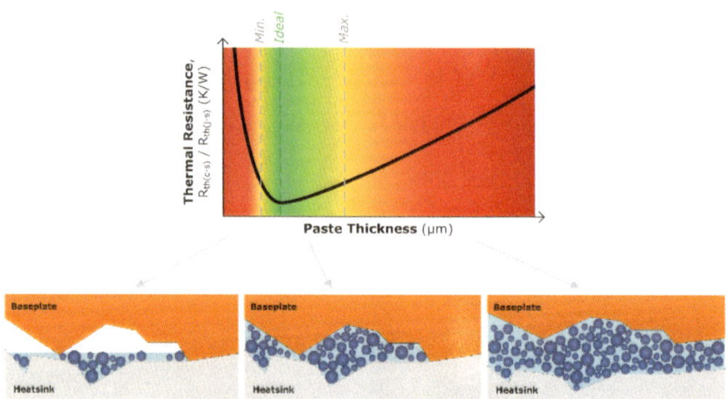

Relationship between paste thickness and thermal resistance of system

Heatsink Specification

The surface finish of the heatsink is a critical part of the interface. Specifications for roughness (Rz), unevenness over distance, and allowable "steps" are given for different module types (baseplate or baseplate-less). The heatsink surface finish requirements are given in the respective "Mounting Instructions" document for the product line in question (e.g. SEMiX, SEMITOP, etc.). These specifications must be indicated on the drawing used for machining the heatsink and should be verified on the finished heatsink prior to use. This can be done using optical inspection methods.

Applying to Module vs. Applying to Heatsink

The decision of whether to apply paste first to the module or to the heatsink is based on the mechanical design of the product and the order assembly during the production process. Generally speaking, a screen/stencil is used to apply paste to the heatsink as it physically easier and screens/stencils can support patterns for multiple modules at once. For this reason, the following screen/stencil examples will show application of the paste to the heatsink, though the processes can be adapted for direct module application.

Preparation of Paste

Most pastes benefit from a rudimentary mixing prior to application to check for homogenization of the carrier and filler materials and inform the operator if anything drastic has occurred to the consistency of stored paste. This can be as simple as stirring the paste within its shipping container prior to dispensing on the production line to as complex as running the materials through an automatic mixer as is used in fully automated screen printing. Well-mixed paste should show uniform color and consistency, without any standing carrier material (i.e. oil). Pastes containing a solvent may require mixing per the manufacturer to ensure that the solvent is evenly distributed.

Roller

For prototyping and low-quantity production, a roller can be used to apply thermal paste directly to the bottom of the module figure below. While low cost, this method results in inconsistent thicknesses and is messy. It does have the advantage that the thickness of the thermal paste layer can be directly measured prior to mounting the module.

A rubber roller (also known as a "brayer" in the art/print industry) with a medium durometer (Shore hardness 50A to 70A) is recommended. This rubber is hard enough that foreign particles should not become easily embedded. The roller should be compatible with paste cleaning agents and not have mechanical features that would allow old paste or dirt to build up. The width of the roller should be narrow enough to fit the baseplate without overhanging but as wide enough to dispense a consistent layer of paste.

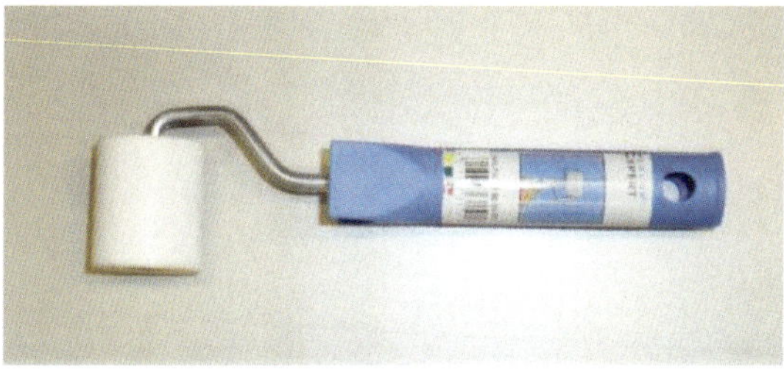

Example roller

Process

1. Paste is dispensed onto a clean, hard, flat surface such as a sheet of glass or hard plastic tray.
2. The roller is run back-and-forth through the paste until a thin layer is spread evenly over the entire roller figure below. This is usually accompanied by high-pitched crackling sound, as the surface tension of the paste is broken as the roller pulls the paste up.

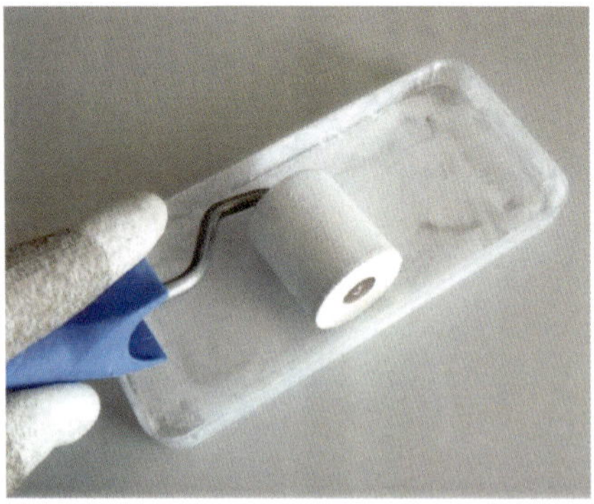

Working the paste onto the roller using a hard plastic tray

3. The roller is applied to the baseplate of the module with light pressure and worked back-and-forth, first in one direction figure below and then in the perpendicular direction (i.e. both X- and Y-axes) until a thin, even layer is present across the entire baseplate.

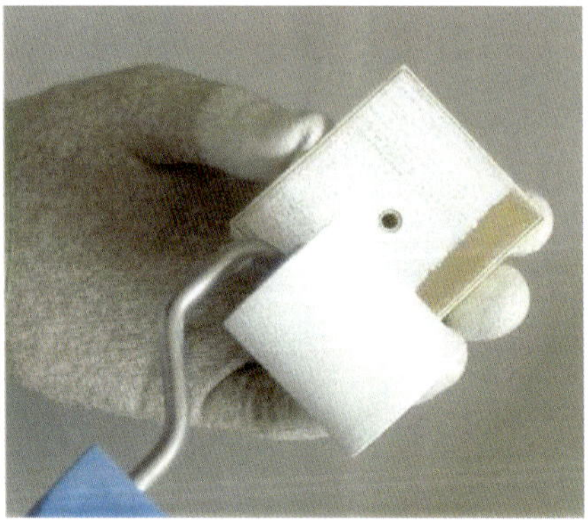

Applying paste to a MiniSKiiP module

4. The layer thickness is checked
5. If too much paste is present on the roller, paste can be removed by running the roller across a "dry" section of the glass sheet.

Screen Printing

Screen printing is a well-established method of printing inks in the fabric and print industries and is well suited for paste application. A fabric mesh is stretched across a frame and a rectangular opening the size of the module baseplate is printed on the mesh using photo-resistive inks. The screen is then pre-wetted with the paste ("flooded"). Finally, the screen is placed a small distance above the heatsink or module and a synthetic rubber scraper ("squeegee") is drawn across the screen with a pre-set angle and pressure. As the screen makes contact with the surface to be printed, the paste is pushed through the mesh and deposited. Since the resulting layer of paste is uniform (without a pattern), the thickness of the paste can be directly measured.

Semi-automated equipment is available that controls the height and position of the screen as well as the pressure applied to the scraper.

Materials

While traditionally known as "silkscreen", modern high-performance screen material is made from plastics such as polyamide yarn ("nylon"), polyethylene terephthalate (PET) or other polyesters and monofilaments. For extra screen longevity, stainless steel mesh is available.

The scraper consists of an extruded aluminum handle to which a medium hard (Shore hardness 60A to 80A) rubber blade has been bolted. Harder rubbers (80A) are used for pastes that are more viscous.

Mesh Count and Thread Size

The most common parameters for defining the mesh are the threads per unit area and the diameter of the thread figure below. The nomenclature varies by measurement unit (cm vs. inches) and material construction (monofilament vs. multifilament) so care must be taken when specifying mesh count depending on the locale (United States vs. Europe). In general, mesh counts for printing thermal paste are much "coarser" than those used for printmaking as the resulting material thickness is greater and less concern is given to feature detail.

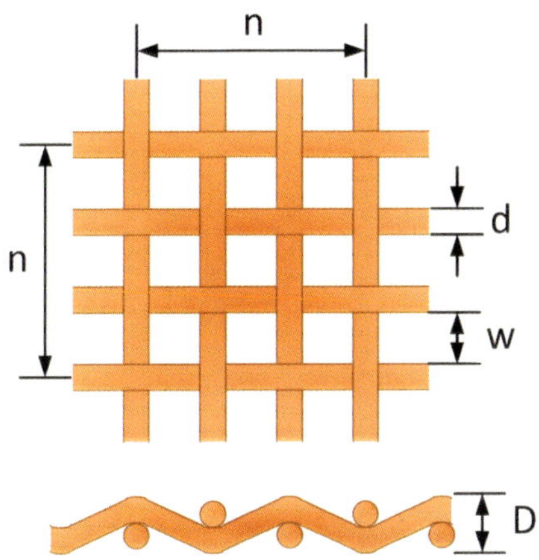

Definition of mesh parameters

Percent open area: $a_0 = \left(\frac{w}{w+d}\right)^3 \cdot 100\%$

Approximate volume of paste: $V_{paste} =\sim 0.75 \cdot D \cdot \left(\frac{w}{w+d}\right)^2$

It has performed some example tests using PET mesh and Wacker P12 thermal paste (table below).

Table: Experimental results

Designation (DIN standard)	77-55	43-80	27-140
Mesh count, n	77 threads/cm2	43 threads/cm2	27 threads/cm2
Thread diameter, d	55µm	80µm	140µm
Resulting paste thickness	20-30µm	40-50µm	70-80µm

Process (Manual)

1. A cleaned heatsink is placed into position under the screen. While heatsinks should be degreased as part of the machining process, it is important to use a lint-free rag with a mild solvent such as isopropyl alcohol to remove fingerprints, dust or other residue just prior to the screen printing process.

Degreased heatsink and lint-free rag

2. A thick line of paste is deposited at the far end of the stencil using a spatula. The screen is "flooded" by drawing the scraper across the screen with light pressure (not making contact with the heatsink).

Flooding of screen (left to right)

3. The screen is lowered into the final printing position (typ. 4-7mm above the heatsink surface depending on screen tension).

Underside of screen showing gap

4. The scraper is drawn across the screen once with high pressure, pressing the screen onto the heatsink and depositing a layer of paste.

Application of paste (right to left)

5. The screen is lifted and the heatsink removed and inspected.

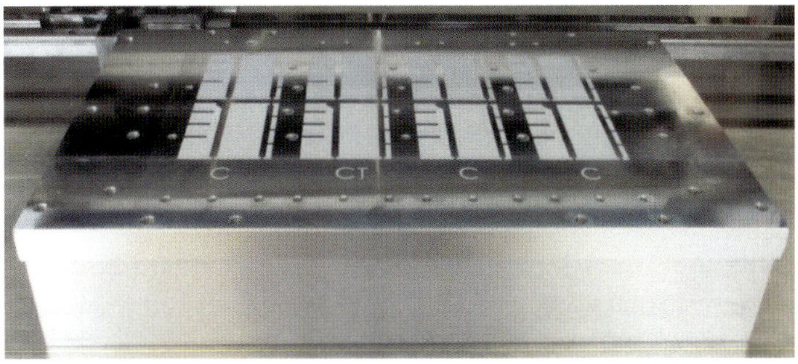

Completed paste pattern

Stencil Printing

The stencil method of paste application is very similar to the screen printing method; however, the fabric screen is replaced with a steel stencil out of which a pattern has been cut figure below. A scraper is used to press the paste into the stencil features, and the use of a pattern ensures that the scraper is parallel to the heatsink at all points. The pattern remains on the surface until the module has been pressed down and temperature cycled, at which point the paste flows to fill voids. The shape, size, and spacing of the holes along with the thickness of the stencil determine how thick the resulting paste is once the module is mounted. Therefore, it is not possible to directly measure the applied thickness of the paste prior to mounting the module. The stencil printing method gives some advantages, including:

a. Ease of design and specification of the stencil for individuals not familiar with the screen printing industry.

b. Allowance for the use of complex thermal paste distributions optimized for chip location and baseplate curvature (typically only performed by the module manufacturer when supplying preapplied TIM).

c. Equipment longevity as the stencil is typically more wear-resistant than a fabric screen.

Stainless steel stencil and resulting paste pattern on bottom of module

Stencil Design

Due to its strength and solvent resistance, stainless steel is usually used for the stencil material. The pattern is cut in the steel using a laser or other process that results in a clean edge on the stencil geometry. Processes such as stamping should be avoided as they can leave a rolled edge or sharp feature that may cause the stencil not to sit flush or cause scratches in the heatsink surface.

The feature size of the stencil should be small enough that the straightness of the scraper edge does not influence the thickness of the paste (i.e. wide openings should be avoided). A honeycomb pattern is most common as it gives an even distribution and can easily be specified on a drawing using two dimensions figure below.

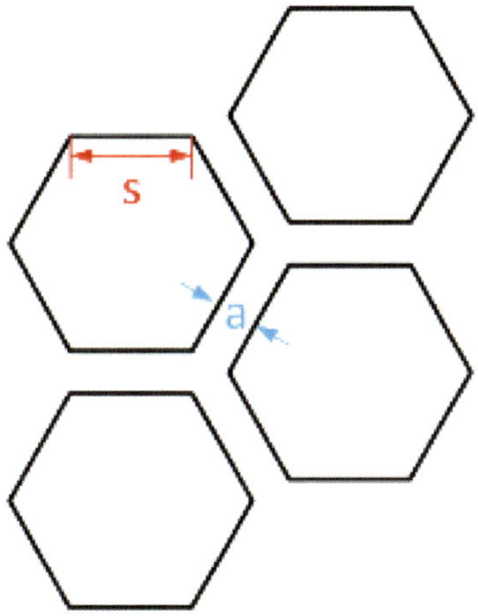

Specification of honeycomb pattern

The "correct" honeycomb pattern to achieve a target past thickness varies by stencil thickness, paste type, and module size (table below).

Table: Example honeycomb parameters for MiniSKiiP modules

Module Type	Honeycomb Edge Length, "s" (mm)	Honeycomb Pitch, "a" (mm)	Stencil Thickness (µm)	Nominal Paste Thickness, Wacker P12 (µm)
MiniSKiiP 1	0.75	0.7	75	30
MiniSKiiP 2	1.1	0.5	100	55
MiniSKiiP 3	0.85	0.55	100	40

Process (Manual)

The overall stencil printing process is very similar to the screen printing process with two notable differences:

a. The stencil is placed in contact with the surface to be stenciled.

b. The pattern is printed in one pass with the scraper (no "flooding" of the stencil is required).

The general steps for stencil printing are then:

1. A cleaned heatsink is placed into position under the stencil.
2. The stencil is lowered into position, making contact with the heatsink. A line of paste is applied to one end of the stencil in front of the scrape using a spatula.
3. The scraper is drawn across the stencil once with high pressure, pressing the paste into the stencil features.
4. The stencil is lifted and the heatsink removed and inspected.

Line of paste deposited prior to drawing scraper across stencil

Screen vs. Stencil Application

The question of whether to use a stencil or screen depends on the required thickness of the paste. Metal stencils thinner than 60μm are not common so for thinner paste thicknesses a screen may be more suitable. For example, in order to reach an effective thickness of less than 20μm, the distance "a" between openings needs to be increased in order to reach the same thickness as can be achieved by a thinner screen. This increases the risk of inhomogeneous paste distribution. Figure below shows a hypothetical comparison where a honeycomb pattern is applied to a silkscreen and compared with the equivalent pattern on a stencil required to achieve a target thickness. As noted previously, screen printing normally does not require such a pattern.

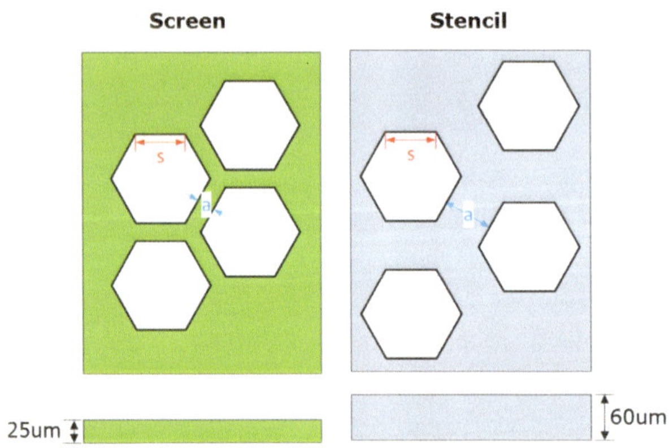

Hypothetical comparison of required screen and stencil patterns to achieve the same paste thickness

In addition, when the module requires a flexible material in order to print the paste (due to mechanical design) the screen is preferred.

Screen and Stencil Wear

Like any other tool, both screens and stencils will wear out over time due to repeated abrasion. As part of a Statistical Process Control (SPC) in a production process, resulting paste thickness should be regularly evaluated and recorded. If it is noticed that the resulting paste thickness is varying or that paste patterns are no longer sharp, the screens/stencils should be examined for wear and replaced if necessary.

Rework and Clean-Up

Paste that has been misapplied should be removed completely prior to reapplication. Most pastes are soluble with petroleum-derived solvents (e.g. white spirits, toluene, kerosene, etc.) though the exact recommended solvents are given on the manufacturers' datasheets. However, isopropyl alcohol is commonly used as it is less toxic and leaves little residue. As it is already commonly used in the electronics industry, it is recommended for both cleaning a heatsink prior to installation as well as removing thermal paste.

Use of a lint-free, disposable rag is strongly recommended to avoid introducing any particles that may get between the module and heatsink.

For screens and stencils, automated washing machines are available. If hand cleaning is performed, care must be taken not to bend or distort the patterns. Screens/stencils are often lightly scrubbed with a solvent and cleaned with compressed air.

Storage and Handling of Paste

Thermal paste does have a shelf life, after which the filler and carrier materials may become so separated that they are unable to be effectively re-mixed for application. Like any other product, the shelf life is also affected by the storage conditions (e.g. temperature, humidity). Paste manufacturers should be able to provide shelf life (e.g. 5 years for a standard silicone type paste) and storage conditions (i.e. temperature and relative humidity).

Once paste has been introduced to the assembly line, contamination of the paste from foreign objects (dust, dirt, particles) becomes a concern. Therefore it is strongly recommended that a plan be instituted for changing the paste regularly (i.e. per production run or shift) and the tooling thoroughly cleaned using the recommended solvents.

Biot Number

The Biot Number is a dimensionless group named after J. B. Biot who, in 1804, analysed the interaction between conduction in a solid and convection at its surface. The numerical value of Biot Number (Bi) is a criterion which gives a direct indication of the relative importance of conduction and convection in determining the temperature history of a body being heated or cooled by convection at its surface. Bi should always be enumerated at the outset to identify transient conduction problems which may be treated simply as lumped parameter problems, for which Bi < 0.1 and for which it is seldom necessary to solve the conduction equation, i.e., convection is the rate controlling process.

Figure below may be used to impute physical meaning to Bi, and shows the limiting steady-state temperature distributions in two plates [(a) high λ and (b) low λ] which have cooled from an initial uniform temperature T_1 as a result of exposure to a cooling flow at T_2 along one face.

For both plates, a heat balance at the cooled surface can be constructed by using Fourier's Law for the conduction flux in the solid at the surface and Newton's Law of Cooling for the convective loss at the surface:

$$\frac{\lambda}{L}(T_1 - T_2) = \alpha(T_{s-}T_2).$$

This heat balance can be rearranged to produce the dimensionless Bi:

$$\left(\frac{T_1 - T_s}{T_s - T_2}\right) = \frac{L/\lambda}{1/\alpha} = \frac{R_{cord}}{R_{conv}} = \frac{\alpha L}{\lambda} = Bi,$$

from which it can be seen that Bi may be considered to be the ratio of the resistance to heat transfer presented by the conduction R_{cond} and the convection R_{conv} processes.

Figure below exemplifies the temperature distribution in a low Bi system (e.g., a steel plate (λ = 35 W/mK) of 5 cm thickness cooling in air [(α = 10 W/m²K) giving Bi = 0.0023)] where, because the resistance to heat flow within the solid is small relative to the resistance presented by the convection processes at the surface, the internal temperature distribution in the solid is relatively uniform. Generally, in bodies of simple geometry, e.g., plates, cylinders, spheres, the error introduced by the assumption of uniform body temperature will be less than 5% when the internal resistance is less than 10% of the external resistance, i.e., when the Bi < 0.1.

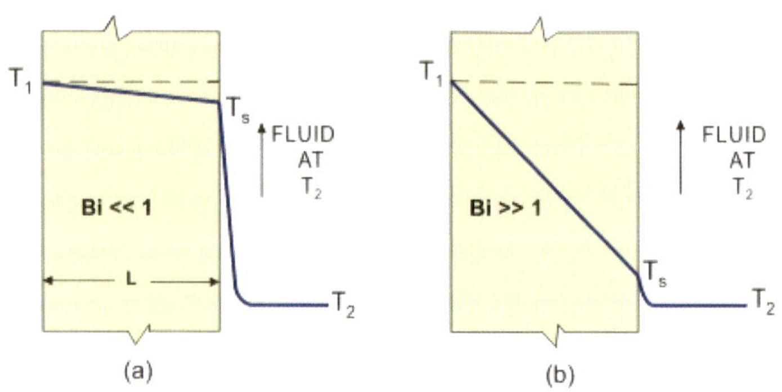

Comparison of fully developed temperature profiles in two plates cooled by the same fluid

Applications

Values of the Biot number smaller than 0.1 imply that the heat conduction inside the body is much faster than the heat convection away from its surface, and temperature gradients are negligible inside of it. This can indicate the applicability (or inapplicability) of certain methods of solving transient heat transfer problems. For example, a Biot number less than 0.1 typically indicates less than 5% error will be present when assuming a lumped-capacitance model of transient heat transfer (also called lumped system analysis). Typically this type of analysis leads to simple exponential heating or cooling behavior ("Newtonian" cooling or heating) since the amount of thermal energy (loosely, amount of "heat") in the body is directly proportional to its temperature, which in turn determines the rate of heat transfer into or out of it. This leads to a simple first-order differential equation which describes heat transfer in these systems.

Having a Biot number smaller than 0.1 labels a substance as "thermally thin," and temperature can be assumed to be constant throughout the material's volume. The opposite is also true: A Biot number greater than 0.1 (a "thermally thick" substance) indicates that one cannot make this assumption, and more complicated heat transfer equations for "transient heat conduction" will be required to describe the time-varying and non-spatially-uniform temperature field within the material body. Analytic methods for handling these problems, which may exist for simple geometric shapes and uniform material thermal conductivity, are described in the article on the heat equation. Examples of verified analytic solutions along with precise numerical values are available. Often such problems are too difficult to be done except numerically, with the use of a computer model of heat transfer.

Together with the Fourier number, the Biot number can be used in transient conduction problems in a lumped parameter solution, which can be written as,

$$\frac{T - T_\infty}{T_0 - T_\infty} = e^{-BiFo}$$

Mass Transfer Analogue

An analogous version of the Biot number (usually called the "mass transfer Biot number", or Bi_m) is also used in mass diffusion processes:

$$Bi_m = \frac{h_m L_C}{D_{AB}}$$

where:

- h_m - film mass transfer coefficient

- L_C - characteristic length
- D_{AB} - mass diffusivity.

Logarithmic Mean Temperature Difference

Log Mean Temperature Difference or LMTD is the driving force for the amount of exchanged heat by a heat exchanger. LMTD approach is quite straight forward and simple. But this approach cannot be used for the cases, where phase change occurs in the heat exchanger. Also, if the specific heats of the fluids change then the LMTD approach cannot be used.

LMTD is the logarithmic mean of temperature difference of the fluids at both the sides of the heat exchangers.

Say, "A" and "B" are the two sides of a heat exchanger, in which fluid "1" and "2" are flowing. Also, let's assume the temperature of the fluid "1" at side "A" and "B" are "Ta1" and "Tb1" respectively and that of fluid "2" are "Ta2" and "Tb2". Then the LMTD (LM) can be calculated as:

$$LM = [ABS(T_{a1} - T_{a2}) - ABS(T_{b1} - T_{b2})] / [LN\{ABS(T_{a1} - T_{a2}) / ABS(T_{b1} - T_{b2})\}]$$

"ABS(Ta1 – Ta2)" signifies the absolute temperature difference between Ta1 and Ta2.

The LMTD is a single value for the whole heat exchanger.

Significance of the LMTD

The LMTD is the driven force for the heat exchange between the two fluids. As the LMTD value increases, the amounts of heat transfer between the two fluids also increase. The LMTD value is used for calculating the heat duty of the heat exchanger. The formula is:

Q = U * A * LMTD

Where,

Q – Heat duty of the heat exchanger (in *watts*)

U – Heat transfer co-efficient (in *watts/Kelvin/Meter square*)

A – Heat transfer area (in meter square)

From the above equation it is clear that the amount of the exchanged heat (Q) is directly proportional to the value of LMTD.

Where ΔT_m is the log mean temperature difference given as

$$\Delta T_m = \frac{\Delta T_1 - \Delta T_2}{\ln\left(\frac{\Delta T_1}{\Delta T_2}\right)}$$

In Parallel Flow Heat Exchanger

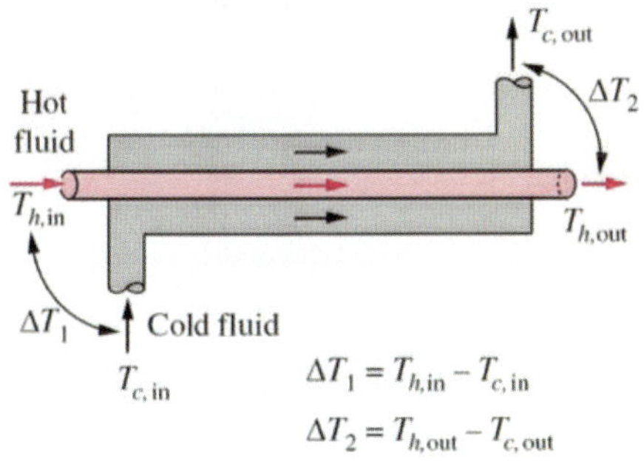

Parallel Flow Heat Exchanger

$$\Delta T_1 = T_{hot,in} - T_{cold,in}$$
$$\Delta T_2 = T_{hot,out} - T_{cold,out}$$

In Counter Flow Heat Exchanger

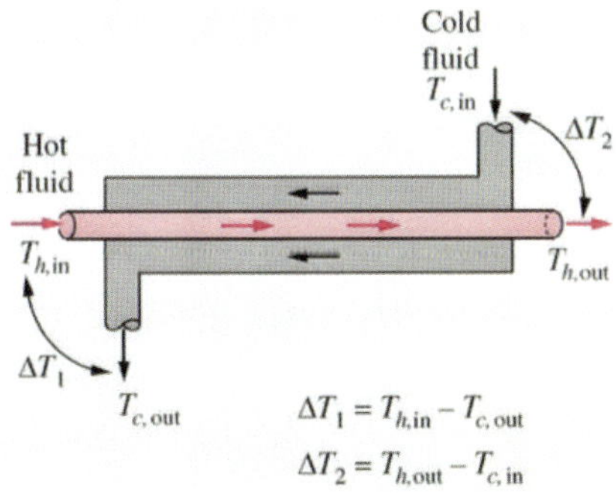

Counter Flow Heat Exchanger

$$\Delta T_1 = T_{hot,in} - T_{cold,out}$$
$$\Delta T_2 = T_{hot,out} - T_{cold,in}$$

Assumptions Made in LMTD Method

1. The overall heat transfer coefficient (U) is constant throughout the heat energy
2. The specific heats (C_p) and mass flow rates of both fluids is constant.

3. The flow conditions are steady.
4. There is no change of phase either of the fluid during the heat transfer.
5. There is no loss of heat to the surroundings, due to heat exchanger being perfectly insulated.
6. Axial conduction along the tubes of heat exchanger is negligible.
7. The changes in kinetic energy and potential energy are negligible.

Derivation

Assume heat transfer is occurring in a heat exchanger along an axis *z*, from generic coordinate *A* to *B*, between two fluids, identified as *1* and *2*, whose temperatures along *z* are $T_1(z)$ and $T_2(z)$.

The local exchanged heat flux at *z* is proportional to the temperature difference:

$$q(z) = U(T_2(z) - T_1(z)) / D = U(\Delta T(z)) / D,$$

where *D* is the distance between the two fluids.

The heat that leaves the fluids causes a temperature gradient according to Fourier's law:

$$\frac{dT_1}{dz} = k_a(T_1(z) - T_2(z)) = -k_a \Delta T(z)$$

$$\frac{dT_2}{dz} = k_b(T_2(z) - T_1(z)) = k_b \Delta T(z)$$

where k_a and k_b are the thermal conductivities of the intervening material at points A and B respectively. Summed together, this becomes

$$\frac{d\Delta T}{dz} = \frac{d(T_2 - T_1)}{dz} = \frac{dT_2}{dz} - \frac{dT_1}{dz} = K\Delta T(z)$$

where $K = k_a + k_b$.

The total exchanged energy is found by integrating the local heat transfer *q* from *A* to *B*:

$$Q = \int_A^B q(z)dz = \frac{U}{D}\int_A^B \Delta T(z)dz = \frac{U}{D}\int_A^B \Delta T\, dz$$

Use the fact that the heat exchanger area *Ar* is the pipe length *B-A* multiplied by the interpipe distance *D*:

$$Q = \frac{UAr}{(B-A)}\int_A^B \Delta T\, dz = \frac{UAr\int_A^B \Delta T\, dz}{\int_A^B dz}$$

In both integrals, make a change of variables from z to ΔT:

$$Q = \frac{UAr\int_{\Delta T(A)}^{\Delta T(B)} \Delta T \frac{dz}{d\Delta T} d(\Delta T)}{\int_{\Delta T(A)}^{\Delta T(B)} \frac{dz}{d\Delta T} d(\Delta T)}$$

With the relation for ΔT found above, this becomes

$$Q = U \times Ar \times \frac{\Delta T(B) - \Delta T(A)}{\ln[\Delta T(B)/\Delta T(A)]},$$

Integration at this point is trivial, and finally gives:

$$Q = U \times Ar \times \frac{\Delta T(B) - \Delta T(A)}{\ln[\Delta T(B)/\Delta T(A)]},$$

from which the definition of LMTD follows.

LMTD Correction Factor

We have seen for co-current or counter current flow system. The average driving force for heat transfer was defined by log mean temperature difference (LMTD). Thus the LMTD can be used for 1-1 exchangers for co-current and counter current. However, for multi pass exchangers (1-2, 2-4, etc.) the fluids are not always in co-current or counter current flow. The deviation for co-current or counter current flow causes a change in the average driving force. Therefore, in order to use true heat transfer driving force, a correction factor is required into the LMTD. Thus, the heat transfer rate can be written as given below,

$$q = U_d A(F_T \Delta T_m)$$

where,

U_d = overall heat transfer coefficient including fouling/dirt

A = heat transfer area

$F_T \Delta T_m$ = true average temperature difference.

F_T = LMTD correction factor

It is to be noted that the following assumption have been considered for developing LMTD,

1. The overall heat transfer coefficient is constant throughout the exchanger
2. In case any fluid undergoes for phase change (e.g., in condenser), the phase change occurs throughout the heat exchanger and the constant fluid temperature prevails throughout the exchanger.
3. The specific heat and mass flow rate and hence the heat capacity rate, of each fluid is constant.
4. No heat is lost in to the surroundings.
5. There is no conduction in the direction of flow neither in the fluids nor in the tube or shell walls.
6. Each of the fluids may be characterized by a single temperature, at any cross section in the heat exchanger that is ideal transverse mixing in each fluid is presumed.

F_T, the LMTD correction factor can be directly obtained from available charts in the literature. These charts were prepared from the results obtained theoretically by solving the temperature distribution in multi-pass heat exchangers.

Figures given below show the two generally used heat exchangers and their corresponding plots for finding F_T. It may be noted that the given figures have the representative plots and any standard book on heat transfer may be consulted for the accurate results.

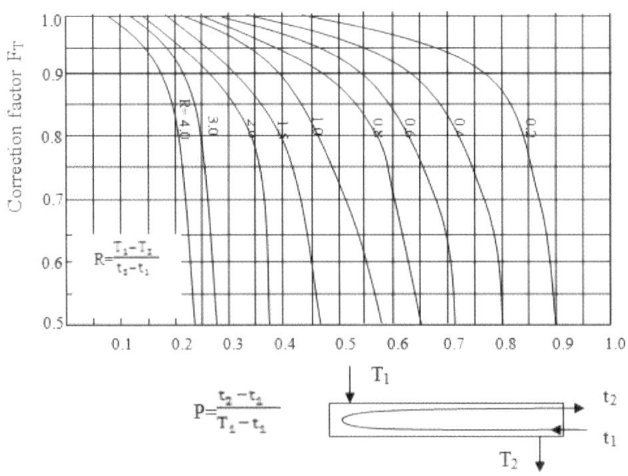

F_T plot for 1-2 exchanger; t: cold fluid in the tube; T: hot fluid in the shell; 1: inlet; 2: outlet

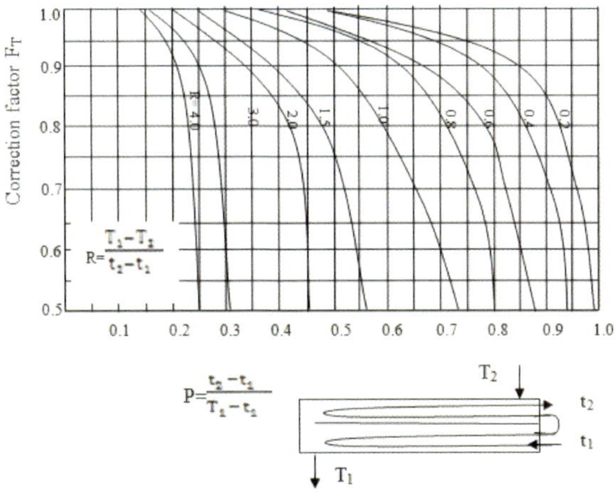

F_T plot for 2-4 exchanger; t: cold fluid in the tube; T: hot fluid in the shell; 1: inlet; 2: outlet

It should be noted that in case of condensation or evaporation the correction factor becomes unity (F_T =1). While designing a heat exchanger, the rule of thumb is that the F_T should not be less than 0.8.

References

- Cole, K.D.; Beck, J.V.; Haji-Sheikh, A.; Litkouhi, B. (2011), Heat Conduction Using Green's Functions (2nd ed.), CRC Press, ISBN 978-1-43-981354-6
- Perona, P; Malik, J. (1990), "Scale-Space and Edge Detection Using Anisotropic Diffusion", IEEE Transactions on Pattern Analysis and Machine Intelligence, 12 (7): 629–639, doi:10.1109/34.56205
- Sam Zhang; Dongliang Zhao (19 November 2012). Aeronautical and Aerospace Materials Handbook. CRC Press. pp. 304–. ISBN 978-1-4398-7329-8. Retrieved 7 May 2013
- Einstein, Albert (1905), "Über die von der molekularkinetischen Theorie der Wärme geforderte Bewegung von in ruhenden Flüssigkeiten suspendierten Teilchen", Annalen der Physik, 322 (8): 549–560, Bibcode:1905AnP...322..549E, doi:10.1002/andp.19053220806
- Incropera, Frank P.; David P. DeWitt; Theodore L. Bergman; Adrienne S. Lavine (2007). Fundamentals of Heat and Mass Transfer (6th ed.). John Wiley & Sons. pp. 260–261. ISBN 978-0-471-45728-2

Convection: Heat Transfer

Convection is the heat transfer in fluids that occurs due to bulk transfer or movement of molecules within the fluid. Convection can be natural, forced or mixed where the two occur in combination. This chapter closely examines the key concepts of convection and its different mechanisms to provide an extensive understanding of heat transfer by convection.

Convection

Convection is the flow of heat through a bulk, macroscopic movement of matter from a hot region to a cool region, as opposed to the microscopic transfer of heat between atoms involved with conduction. Suppose we consider heating up a local region of air. As this air heats, the molecules spread out, causing this region to become less dense than the surrounding, unheated air. being less dense than the surrounding cooler air, the hot air will subsequently rise due to buoyant forces - this movement of hot air into a cooler region is then said to transfer heat by convection.

Heating a pot of water on a stove is a good example of the transfer of heat by convection. When the stove is first turned on heat is transferred first by conduction between the element through the bottom of the pot to the water. However, eventually the water starts bubbling - these bubbles are actually local regions of hot water rising to the surface, thereby transferring heat from the hot water at the bottom to the cooler water at the top by convection. At the same time, the cooler, more dense water at the top will sink to the bottom, where it is subsequently heated. These convection currents are illustrated in the following figure.

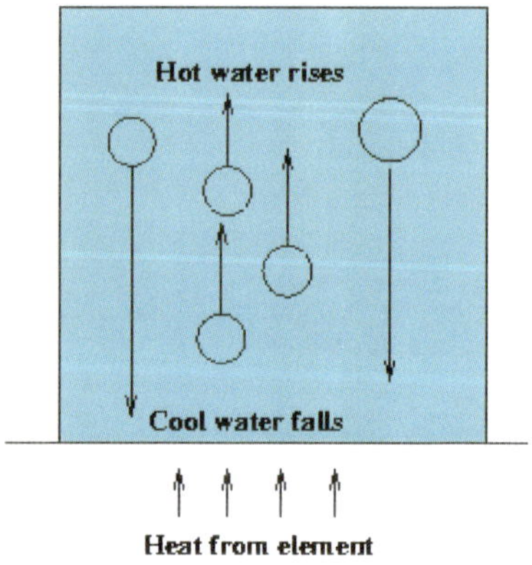

Convection currents in boiling water

Consider now two regions separated by a barrier, one at a higher pressure relative to the other, and subsequently remove the barrier, as in the following figure. These convection currents are illustrated in the following figure.

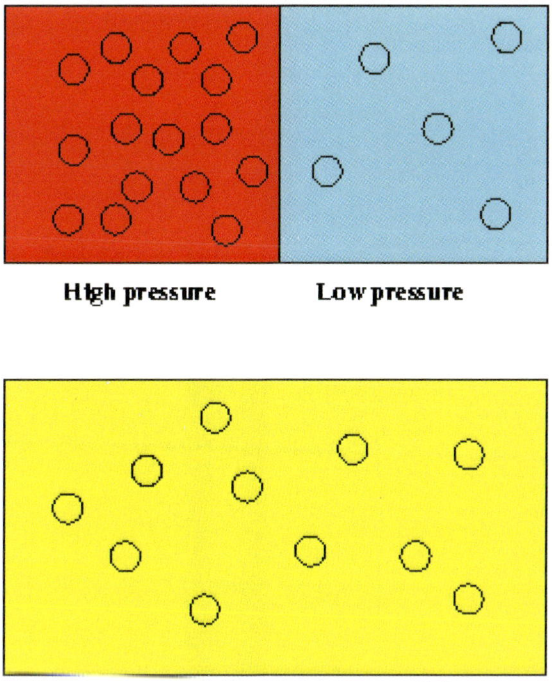

Flow of material through a pressure difference

When the barrier is removed, material in the high pressure (high density) area will flow to the low pressure (low density) area. If the low pressure region was originally created by heating of the material, one sees that movement of material in this way is an example of heat flow by convection.

An important example of convection currents that can be interpreted in this manner is the creation of breezes over land masses next to large bodies of water. Water has a larger heat capacity than land, and subsequently holds heat better. It therefore takes longer to change its temperature, either upward or downward. Thus, during the day the air above the water will be cooler than that over the land. This creates a low pressure area over the land, relative to the high pressure area over the water, and subsequently one finds breezes blowing from the water to the land. On the other hand, during the night water cools off more slowly than the land, and the air above the water is slightly warmer than over the land. This creates a low pressure area over the water relative to the high pressure area over the land, and breezes will blow from the land to the water. These are illustrated in the following figure.

Convection can arise spontaneously (or naturally or freely) through the creation of convection cells or can be forced by propelling the fluid across the object or by the object through the fluid.

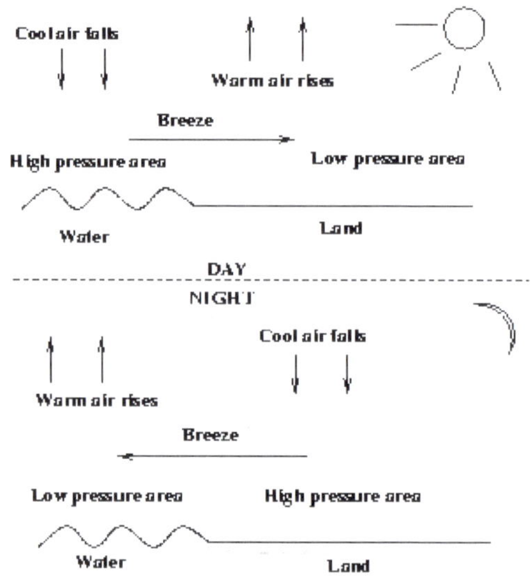

Formation of breezes near large bodies of water

Spontaneous Convection

Spontaneous convection is driven by buoyancy for the most part and surface tension to a lesser extent.

- exposed surface area
- viscosity
- density
- conductivity (when conductivity is high, there is no need for convection)
- acceleration due to gravity

Examples

- General
  - boyancy driven convection — bénard cells — Rayleigh-Bénard convection
  - surface tension driven convection — surface downwelling — Marangoni effect — Rayleigh-Bénard-Marangoni convection
- Atmospheric circulation
  - local
    - updrafts — anabatic winds
      - thermals
      - cumulus clouds: stratoculumus, altocumulus, cirrocumulus, and big daddy cumulonimbus

        - downdrafts -- katabatic winds
            - downbursts, microbursts
            - mountain winds: chinook, foehn (föhn), santa ana, diablo, viento zonda, bergwind
    - global — hadley cells (tropical), ferrel cells (temperte - less well defined), polar cells (polar, obviously)
        - polar high
        - polar easterlies
        - polar front
        - prevailing westerlies — temperate zone — zone of mixing
        - horse lattitudes — subtropical deserts
        - trade winds (ex. alizé)
        - doldrums — intertropical convergence zone
            - ITCZ follows the sun (thermal equator)
            - Monsoons
- Ocean currents are driven by a combination of temperature and salinity gradients (thermohaline circulation) in the deep ocean, by winds near the surface, and by topography everywhere water touches land
    - gulf stream
        - the historical first reported ocean current
        - keeps eurpoe warmer than north america at the same lattitude
    - deep ocean return current and its effect on global climate
        - submarine rivers
        - ocean's conveyor belt
        - affected by salinity of surface waters near Greenland, lower salinity means lower density and less likely to sink, the end of the world is nigh
- Geologic
    - mantle convection drives plate techtonics
    - outer core convection (along with charge separation) generates earth's magnetic field

- Solar
    - core — where the action takes place
    - radiative zone — the sun is a good enough conductor (high density, high conductivity) that energy flows by radiation (why not conduction?)
    - convection zone
    - photosphere — convection cells show up as granules (granule implies small, but small is a relative word)
    - corona — material is thrown out gradually in the solar wind or explosively in solar prominances or coronal mass ejections

Forced Convection

The heat loss due to the forced convection is given by the following equation called Newton's Law of Cooling.

$$P = \frac{dQ}{dt} = hA\,(T - T_0)$$

where,

$P = dQ/dt$ is rate at which heat is transferred

h = convection heat-transfer coefficient (or film coefficient or film conductance)

A = exposed surface area

T = temperature of the immersed object

T_0 = temperature of convecting fluid

There is currently no general theoretical model for analyzing forced convection problems. The heat-transfer coefficients h are can only be described by equations based on empirical analysis. For example, the h of air is approximately equal to...

$$h = 10.45 - v + 10\sqrt{v}$$

where v is the relative speed of the object through the air; that is, the speed of the object through the air or the speed of the air around the object. This equation is valid for speed from 2 to 20 m/s.

Gravitational or Buoyant Convection

Gravitational convection is a type of natural convection induced by buoyancy variations resulting from material properties other than temperature. Typically this is caused by a variable composition of the fluid. If the varying property is a concentration gradient, it is known as solutal convection. For example, gravitational convection can be seen in the diffusion of a source of dry salt downward into wet soil due to the buoyancy of fresh water in saline.

Variable salinity in water and variable water content in air masses are frequent causes of convection in the oceans and atmosphere which do not involve heat, or else involve additional compositional density factors other than the density changes from thermal expansion. Similarly, variable composition within the Earth's interior which has not yet achieved maximal stability and minimal energy (in other words, with densest parts deepest) continues to cause a fraction of the convection of fluid rock and molten metal within the Earth's interior.

Gravitational convection, like natural thermal convection, also requires a g-forceenvironment in order to occur.

Granular Convection

Vibration-induced convection occurs in powders and granulated materials in containers subject to vibration where an axis of vibration is parallel to the force of gravity. When the container accelerates upward, the bottom of the container pushes the entire contents upward. In contrast, when the container accelerates downward, the sides of the container push the adjacent material downward by friction, but the material more remote from the sides is less affected. The net result is a slow circulation of particles downward at the sides, and upward in the middle.

If the container contains particles of different sizes, the downward-moving region at the sides is often narrower than the largest particles. Thus, larger particles tend to become sorted to the top of such a mixture. This is one possible explanation of the Brazil nut effect.

Solid-state Convection in Ice

Ice convection on Pluto is believed to occur in a soft mixture of nitrogen ice and carbon monoxide ice. It has also been proposed for Europa, and other bodies in the outer solar system.

Thermomagnetic Convection

Thermomagnetic convection can occur when an external magnetic field is imposed on a ferrofluid with varying magnetic susceptibility. In the presence of a temperature gradient this results in a nonuniform magnetic body force, which leads to fluid movement. A ferrofluid is a liquid which becomes strongly magnetized in the presence of a magnetic field.

This form of heat transfer can be useful for cases where conventional convection fails to provide adequate heat transfer, e.g., in miniature microscale devices or under reduced gravity conditions.

Capillary Action

Capillary action is a phenomenon where liquid spontaneously rises in a narrow space such as a thin tube, or in porous materials. This effect can cause liquids to flow against the force of gravity. It occurs because of inter-molecular attractive forces between the liquid and solid surrounding surfaces; If the diameter of the tube is sufficiently small, then the combination of surface tension and forces of adhesion between the liquid and container act to lift the liquid.

Marangoni Effect

The Marangoni effect is the convection of fluid along an interface between dissimilar substances because of variations in surface tension. Surface tension can vary because of inhomogeneous composition of the substances or the temperature-dependence of surface tension forces. In the latter case the effect is known as thermo-capillary convection.

A well-known phenomenon exhibiting this type of convection is the "tears of wine".

Weissenberg Effect

The Weissenberg effect is a phenomenon that occurs when a spinning rod is placed into a solution of liquid polymer. Entanglements cause the polymer chains to be drawn towards the rod instead of being thrown outward as would happen with an ordinary fluid (i.e., water).

Combustion

In a zero-gravity environment, there can be no buoyancy forces, and thus no natural (free) convection possible, so flames in many circumstances without gravity smother in their own waste gases. However, flames may be maintained with any type of forced convection (breeze); or (in high oxygen environments in "still" gas environments) entirely from the minimal forced convection that occurs as heat-induced *expansion* (not buoyancy) of gases allows for ventilation of the flame, as waste gases move outward and cool, and fresh high-oxygen gas moves in to take up the low pressure zones created when flame-exhaust water condenses.

Mathematical Models of Convection

Mathematically, convection can be described by the convection–diffusion equation, also known as the generic scalar transport equation.

Quantifying Natural Versus Forced Convection

In cases of mixed convection (natural and forced occurring together) one would often like to know how much of the convection is due to external constraints, such as the fluid velocity in the pump, and how much is due to natural convection occurring in the system.

The relative magnitudes of the Grashof number and the square of the Reynolds number determine which form of convection dominates. If $\mathrm{Gr} / \mathrm{Re}^2 \gg 1$, forced convection may be neglected, whereas if $\mathrm{Gr} / \mathrm{Re}^2 \ll 1$, natural convection may be neglected. If the ratio, known as the Richardson number, is approximately one, then both forced and natural convection need to be taken into account.

Examples and Applications of Convection

Convection occurs on a large scale in atmospheres, oceans, planetary mantles, and it provides the mechanism of heat transfer for a large fraction of the outermost interiors of our sun and all stars. Fluid movement during convection may be invisibly slow, or it may be obvious and rapid, as in a hurricane. On astronomical scales, convection of gas and dust is thought to occur in the accretion disks of black holes, at speeds which may closely approach that of light.

Heat Transfer

A heat sink provides a large surface area for convection to efficiently carry away heat

Convective heat transfer is a mechanism of heat transfer occurring because of bulk motion (observable movement) of fluids. Heat is the entity of interest being advected (carried), and diffused (dispersed). This can be contrasted with conductive heat transfer, which is the transfer of energy by vibrations at a molecular level through a solid or fluid, and radiative heat transfer, the transfer of energy through electromagnetic waves.

Heat is transferred by convection in numerous examples of naturally occurring fluid flow, such as: wind, oceanic currents, and movements within the Earth's mantle.

Convection is also used in engineering practices of homes, industrial processes, cooling of equipment, etc.

The rate of convective heat transfer may be improved by the use of a heat sink, often in conjunction with a fan. For instance, a typical computer CPU will have a purpose-made fan to ensure its operating temperature is kept within tolerable limits.

Convection Cells

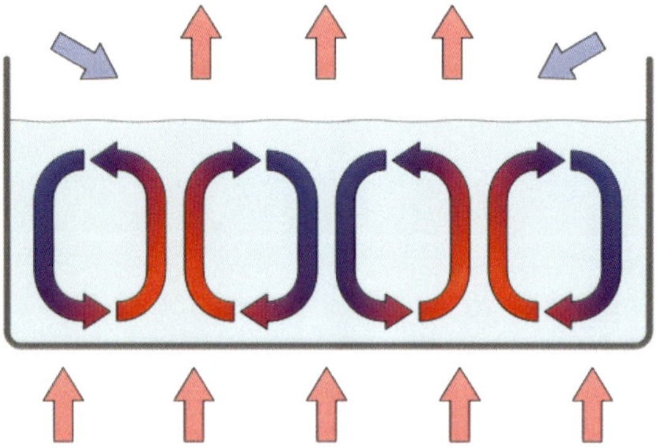

Convection cells in a gravity field

A convection cell, also known as a Bénard cell is a characteristic fluid flow pattern in many convection systems. A rising body of fluid typically loses heat because it encounters a cold surface. In liquid this occurs because it exchanges heat with colder liquid through direct exchange. In the example of the Earth›s atmosphere, this occurs because it radiates heat. Because of this heat loss the fluid becomes denser than the fluid underneath it, which is still rising. Since it cannot descend through the rising fluid, it moves to one side. At some distance, its downward force overcomes the rising force beneath it, and the fluid begins to descend. As it descends, it warms again and the cycle repeats itself.

Atmospheric Circulation

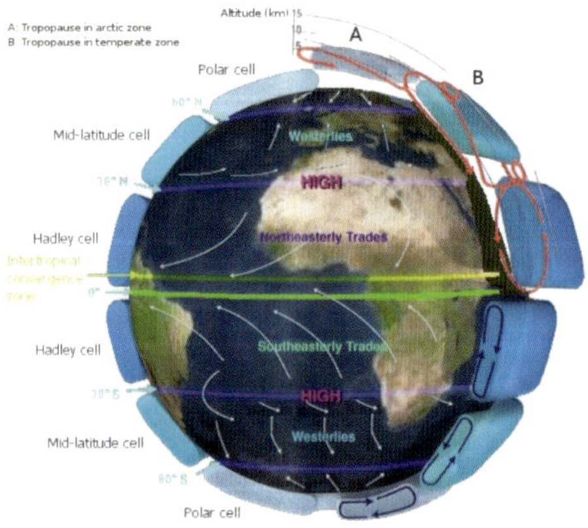

Idealised depiction of the global circulation on Earth

Atmospheric circulation is the large-scale movement of air, and is a means by which thermal energy is distributed on the surface of the Earth, together with the much slower (lagged) ocean circulation system. The large-scale structure of the atmospheric circulation varies from year to year, but the basic climatological structure remains fairly constant.

Latitudinal circulation occurs because incident solar radiation per unit area is highest at the heat equator, and decreases as the latitudeincreases, reaching minima at the poles. It consists of two primary convection cells, the Hadley cell and the polar vortex, with the Hadley cell experiencing stronger convection due to the release of latent heat energy by condensation of water vaporat higher altitudes during cloud formation.

Longitudinal circulation, on the other hand, comes about because the ocean has a higher specific heat capacity than land (and also thermal conductivity, allowing the heat to penetrate further beneath the surface) and thereby absorbs and releases more heat, but the temperature changes less than land. This brings the sea breeze, air cooled by the water, ashore in the day, and carries the land breeze, air cooled by contact with the ground, out to sea during the night. Longitudinal circulation consists of two cells, the Walker circulation and El Niño / Southern Oscillation.

Weather

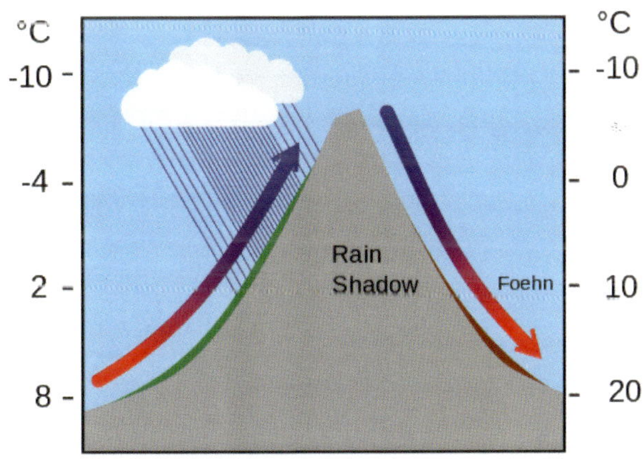

How Foehn is produced

Some more localized phenomena than global atmospheric movement are also due to convection, including wind and some of the hydrologic cycle. For example, a foehn wind is a down-slope wind which occurs on the downwind side of a mountain range. It results from the adiabatic warming of air which has dropped most of its moisture on windward slopes. Because of the different adiabatic lapse rates of moist and dry air, the air on the leeward slopes becomes warmer than at the same height on the windward slopes.

A thermal column (or thermal) is a vertical section of rising air in the lower altitudes of the Earth›s atmosphere. Thermals are created by the uneven heating of the Earth's surface from solar radiation. The Sun warms the ground, which in turn warms the air directly above it. The warmer air expands, becoming less dense than the surrounding air mass, and creating a thermal low. The mass of lighter air rises, and as it does, it cools by expansion at lower air pressures. It stops rising when it has cooled to the same temperature as the surrounding air. Associated with a thermal is a downward flow surrounding the thermal column. The downward moving exterior is caused by colder air being displaced at the top of the thermal. Another convection-driven weather effect is the sea breeze.

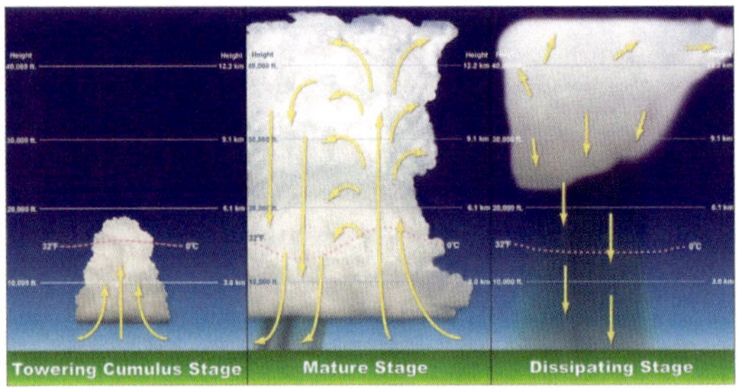

Stages of a thunderstorm's life

Warm air has a lower density than cool air, so warm air rises within cooler air,similar to hot air balloons. Clouds form as relatively warmer air carrying moisture rises within cooler air. As the moist air rises, it cools, causing some of the water vapor in the rising packet of air to condense. When the moisture condenses, it releases energy known as latent heat of condensation which allows the rising packet of air to cool less than its surrounding air, continuing the cloud›s ascension. If enough instability is present in the atmosphere, this process will continue long enough for cumulonimbus clouds to form, which support lightning and thunder. Generally, thunderstorms require three conditions to form: moisture, an unstable airmass, and a lifting force (heat).

All thunderstorms, regardless of type, go through three stages: the developing stage, the mature stage, and the dissipation stage. The average thunderstorm has a 24 km (15 mi) diameter. Depending on the conditions present in the atmosphere, these three stages take an average of 30 minutes to go through.

Oceanic Circulation

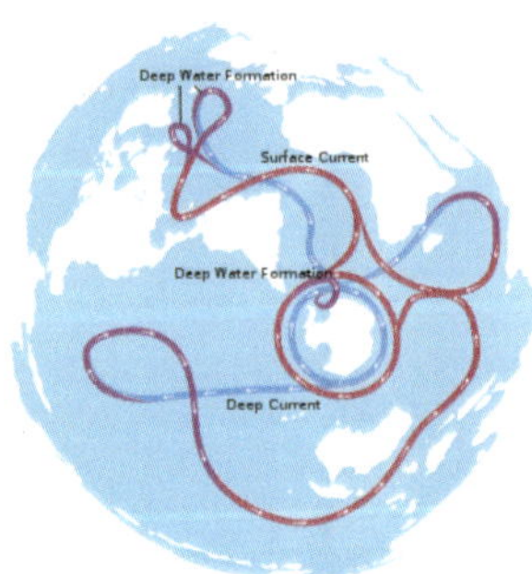

Ocean currents

Solar radiation affects the oceans: warm water from the Equator tends to circulate toward the poles, while cold polar water heads towards the Equator. The surface currents are initially dictated by surface wind conditions. The trade winds blow westward in the tropics, and the westerlies blow eastward at mid-latitudes.This wind pattern applies a stress to the subtropical ocean surface with negative curlacross the Northern Hemisphere, and the reverse across the Southern Hemisphere. The resulting Sverdrup transport is equatorward. Because of conservation of potential vorticity caused by the poleward-moving winds on the subtropical ridge's western periphery and the increased relative vorticity of poleward moving water, transport is balanced by a narrow, accelerating poleward current, which flows along the western boundary of the ocean basin, outweighing the effects of friction with the cold western boundary current which originates from high latitudes. The overall process, known as western intensification, causes currents on the western boundary of an ocean basin to be stronger than those on the eastern boundary.

As it travels poleward, warm water transported by strong warm water current undergoes evaporative cooling. The cooling is wind driven: wind moving over water cools the water and also causes evaporation, leaving a saltier brine. In this process, the water becomes saltier and denser. and decreases in temperature. Once sea ice forms, salts are left out of the ice, a process known as brine exclusion. These two processes produce water that is denser and colder. The water across the northern Atlantic ocean becomes so dense that it begins to sink down through less salty and less dense water. (The convective action is not unlike that of a lava lamp.) This downdraft of heavy, cold and dense water becomes a part of the North Atlantic Deep Water, a southgoing stream.

Mantle Convection

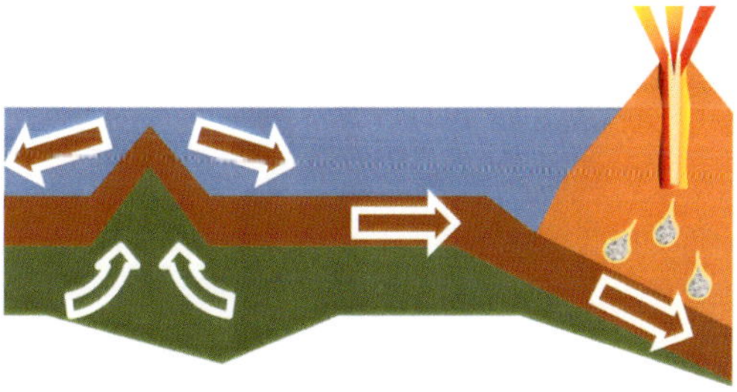

An oceanic plate is added to by upwelling (left) and consumed at a subduction zone (right)

Mantle convection is the slow creeping motion of Earth's rocky mantle caused by convection currents carrying heat from the interior of the earth to the surface. It is one of 3 driving forces that causes tectonic plates to move around the Earth's surface.

The Earth's surface is divided into a number of tectonic plates that are continuously being created and consumed at their opposite plate boundaries. Creation (accretion) occurs as mantle is added to the growing edges of a plate. This hot added material cools down by conduction and convection of heat. At the consumption edges of the plate, the material has thermally contracted to become dense, and it sinks under its own weight in the process of subduction at an ocean trench. This subducted material sinks to some depth in the Earth's interior where it is prohibited from sinking further. The subducted oceanic crust triggers volcanism.

Stack Effect

The Stack effect or chimney effect is the movement of air into and out of buildings, chimneys, flue gas stacks, or other containers due to buoyancy. Buoyancy occurs due to a difference in indoor-to-outdoor air density resulting from temperature and moisture differences. The greater the thermal difference and the height of the structure, the greater the buoyancy force, and thus the stack effect. The stack effect helps drive natural ventilation and infiltration. Some cooling towersoperate on this principle; similarly the solar updraft tower is a proposed device to generate electricity based on the stack effect.

Stellar Physics

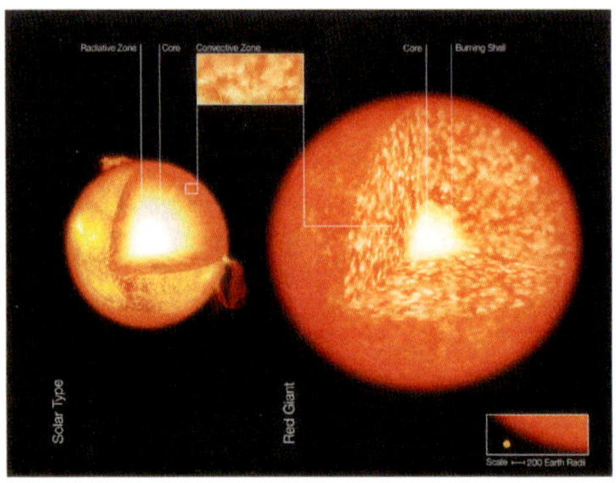

An illustration of the structure of the Sun and a red giant star, showing their convective zones. These are the granular zones in the outer layers of these stars

The convection zone of a star is the range of radii in which energy is transported primarily by convection.

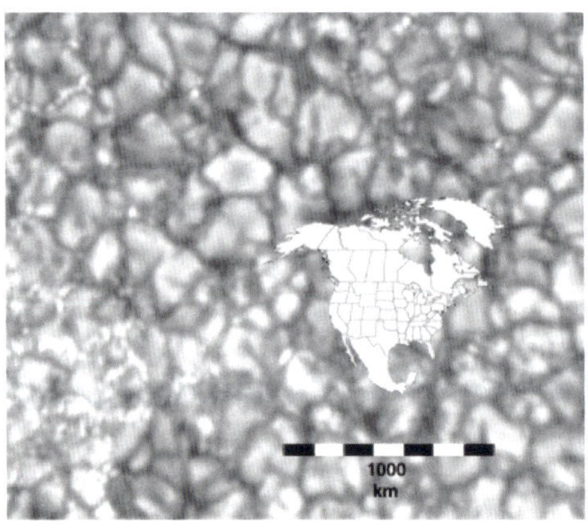

Granules—the tops or upper visible sizes of convection cells, seen on the photosphere of the Sun. These are caused by the convection in the upper photosphere of the Sun. North America is superimposed on the same scale, to indicate scale

Granules on the photosphere of the Sun are the visible tops of convection cells in the photosphere, caused by convection of plasma in the photosphere. The rising part of the granules is located in the center where the plasma is hotter. The outer edge of the granules is darker due to the cooler descending plasma. A typical granule has a diameter on the order of 1,000 kilometers and each lasts 8 to 20 minutes before dissipating. Below the photosphere is a layer of much larger "supergranules" up to 30,000 kilometers in diameter, with lifespans of up to 24 hours.

Natural Convection

Natural convection, known also as free convection is a mechanism, or type of mass and heat transport, in which the fluid motion is generated only by density differencesin the fluid occurring due to temperature gradients, not by any external source (like a pump, fan, suction device, etc.).

In natural convection, fluid surrounding a heat source receives heat and by thermal expansion becomes less dense and rises. Thermal expansion of the fluid plays a crucial role. In other words, heavier (more dense) components will fall, while lighter (less dense) components rise, leading to bulk fluid movement. Natural convection can only occur in a gravitational field or in the presence of another proper acceleration, such as:

- acceleration
- centrifugal force
- Coriolis force

Natural convection essentially does not operate in the orbit of Earth. For example, in the orbiting International Space Station, other heat transfer mechanisms are required to prevent electronic components from overheating.

Similarly as for forced convection, also natural convection heat transfer take place both by thermal diffusion (the random motion of fluid molecules) and by advection, in which matter or heat is transported by the larger-scale motion of currents in the fluid. At the surface, energy flow occurs purely by conduction, even in convection. It is due to the fact, there is always a thin stagnant fluid film layer on the heat transfer surface. But in the next layers both conduction and diffusion-mass movement in the molecular level or macroscopic level occurs. Due to the mass movement the rate of energy transfer is higher. Higher the rate of mass movement, thinner the stagnant fluid film layer will be and higher will be the heat flow rate.

Thermal Expansion

In general, density can be changed by changing either the pressure or the temperature.

Increasing the pressure always increases the density of a material. The effect of pressure on the densities of liquids and solids is very very small. On the other hand, the density of gases is strongly affected by pressure. This is expressed by compressibility. Compressibility is a measure of the relative volume change of a fluid or solid as a response to a pressure change.

The effect of temperature on the densities of liquids and solids is also very important. Most substances expand when heated and contract when cooled. However, the amount of expansion or contraction varies, depending on the material. This phenomenon is known as thermal expansion. The change in volume of a material which undergoes a temperature change is given by following relation:

$$\frac{\Delta V}{V} = a_V \Delta T$$

where ΔT is the change in temperature, V is the original volume, ΔV is the change in volume, and αV is the coefficient of volume expansion.

It must be noted, there are exceptions from this rule. For example, water differs from most liquids in that it becomes less dense as it freezes. It has a maximum of density at 3.98 °C (1000 kg/m³), whereas the density of ice is 917 kg/m³. It differs by about 9% and therefore ice floats on liquid water

Density of Water as a Function of Temperature

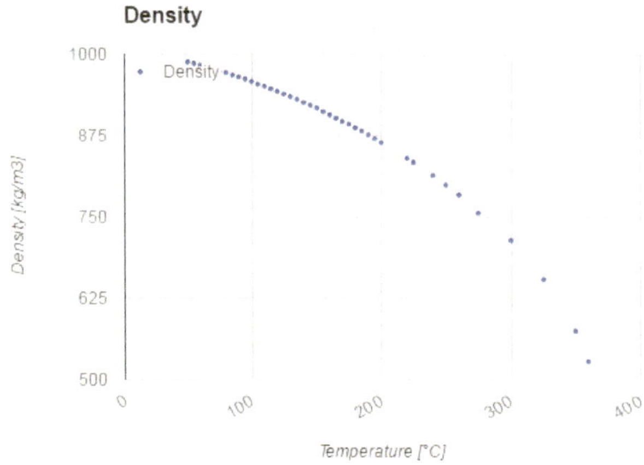

Density of water as a function of temperature

We know that the density of gases and liquids depends on temperature, generally decreasing (due to fluid expansion) with increasing temperature.

The magnitude of the natural convection heat transfer between a surface and a fluid is directly related to the flow rate of the fluid induced by natural convection. The higher the flow rate, the higher the heat transfer rate. The flow rate in case of natural convection is established by the dynamic balance of buoyancy and friction.

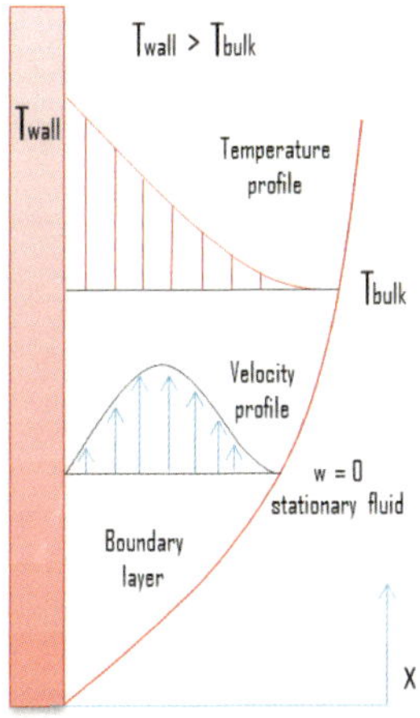

Assume a plate at the temperature T_{wall}, which is immersed in a quiescent fluid at the temperature T_{bulk}, where ($T_{wall} > T_{bulk}$). The fluid close to the plate is less dense than fluid that is further removed. Buoyancy forces therefore induce a natural convection boundary layer in which the heated and lighter fluid rises vertically, entraining heavier fluid from the quiescent region. The resulting velocity distribution is unlike that associated with forced convection boundary layers and depends also on the fluid viscosity. In particular, the velocity is zero at the surface as well as at the boundarydue to viscous forces. It must be noted, a natural convection also develops if ($T_{wall} < T_{bulk}$), but, in this case, fluid motion will be downward.

The presence and magnitude of natural convection also depend on the geometry of the problem. The presence of a fluid density gradient in a gravitational field does not ensure the existence of natural convection currents. This problem is illustrated in the following figure, where a fluid is enclosed by two large, horizontal plates of different temperature ($T_{upper} \neq T_{lower}$).

1. In case A the temperature of the lower plate is higher than the temperature of the upper plate. In this case, the density decreases in the direction of the gravitational force. This geometry induces fluid circulation and heat transfer occurs via natural circulation. The heavier fluid will descend, being warmed in the process, while the lighter fluid will rise, cooling as it moves.

2. In case B the temperature of the lower plate is lower than the temperature of the upper plate. In this case, the density increases in the direction of the gravitational force. This geometry leads to stable conditions, stable temperature gradient and does not induce fluid circulation. Heat transfer occurs solely via thermal conduction.

Since the natural convection is strongly dependent on the geometry, most heat transfer correlations in natural convection are based on experimental measurements and engineers often use proper characteristic numbers to describe natural convection heat transfer.

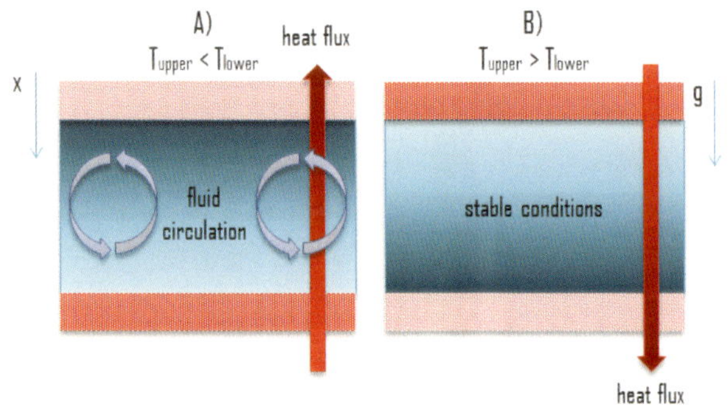

Natural Convection - Laminar and Turbulent

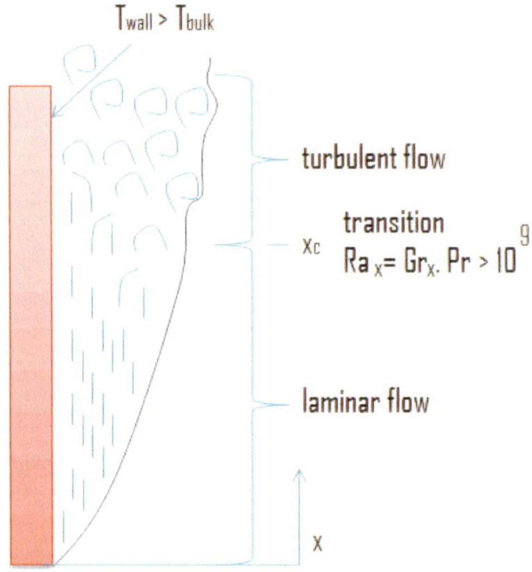

It is important to note that natural convection boundary layers are not restricted to laminar flow. As with forced convection, hydrodynamic instabilities may arise. That is, disturbances in the flow may be amplified, leading to transition from laminar to turbulent flow. For a vertical flat plate, the flow turns turbulent for value of:

$$Ra_x = Gr_x . Pr > 10^9$$

As in forced convection the microscopic nature of flow and convection correlations are distinctly different in the laminar and turbulent regions.

Natural Convection – Correlations

As was written, most heat transfer correlations in natural convection are based on experimental measurements and engineers often use proper characteristic numbers to describe natural convection heat transfer. The characteristic number that describes convective heat transfer (i.e. the heat transfer coefficient) is the Nusselt number, which is defined as the ratio of the thermal energy convected to the fluid to the thermal energy conducted within the fluid. The Nusselt number represents the enhancement of heat transfer through a fluid layer as a result of convection relative to conduction across the same fluid layer. But in case of free convection, heat transfer correlations (for the Nusselt number) are usually expressed in terms of the Rayleigh number.

The Rayleigh number is used to express heat transfer in natural convection. The magnitude of the Rayleigh number is a good indication as to whether the natural convection boundary layer is laminar or turbulent. The simple empirical correlations for the average Nusselt number, Nu, in natural convection are of the form:

$$Nu_x = C. Ra_x^n$$

The values of the constants C and n depend on the geometry of the surface and the flow regime, which is characterized by the range of the Rayleigh number. The value of n is usually n = 1/4 for laminar flow and n = 1/3 for turbulent flow.

For example:

Correlation: Vertical plate – laminar	Validity:
$Nu_L = 0.59 Ra_L^{1/4}$	$10^4 \leq Ra_L \leq 10^9$
Correlation: Vertical plate – turbulent	Validity:
$Nu_L = 0.1 Ra_L^{1/3}$	$10^9 \leq Ra_L \leq 10^{13}$

Example: Natural Convection – Flat Plate

A 10cm high vertical plate is maintained at 261°C in 260°C compressed water (16MPa). Determine the Nusselt number using the simple correlation for a vertical flat plate.

Correlation : Vertical plate – turbulent	Validity
$Nu_L = 0.1 Ra_L^{1/3}$	$10^9 \leq Ra_L \leq 10^{13}$

To calculate the Rayleigh number, we have to know:

- the coefficient of thermal expansion, which is: $\beta = 0.0022$
- the Prandtl number (for 260°C), which is: Pr = 0.87
- the kinematic viscosity (for 260°C), which is $\nu = 0.13 \times 10^{-6}$ (note that, this value is significantly lower than that for 20°C)

The resulting Rayleigh number is:

$$Ra_L = Gr_L . \mathrm{Pr} = \frac{g\beta(T_{wall} - T_\infty)L^3}{\nu^2} . \mathrm{Pr}$$

$$Ra_L = \frac{9.81 x 0.0022 x 1 x 0.1^3}{(0.13 x 10^{-6})^2} x 0.87 = 1.11 x 10^9$$

The resulting Nusselt number, which represents the enhancement of heat transfer through a fluid layer as a result of convection relative to conduction across the same fluid layer is:

$$Nu_L = 0.1 Ra_L^{1/3} = 0.1 \, x \, (1.11 \, x \, 10^9)^{1/3} = 102.8$$

Forced Convection

Forced convection is a special type of heat transfer in which fluids are forced to move, in order to increase the heat transfer. This forcing can be done with a ceiling fan, a pump, suction device, or other.

Many people are familiar with the statement that "heat rises". This is a simplification of the idea that hot fluids are almost always less dense than the same fluid when cold, but there are exceptions. This difference in density makes hotter material naturally end up on top of cooler material due to the higher buoyancy of the hotter material.

Natural convection can create a noticeable difference in temperature within a home. Often this becomes places where certain parts of the house are warmer and certain parts are cooler. Forced convection creates a more uniform and therefore comfortable temperature throughout the entire home. This reduces *cold spots* in the house, reducing the need to crank the thermostat to a higher temperature, or putting on sweaters.

Convection is a heat transfer mechanism where heat moves from one place to another through fluid currents. Forced convection is simply using this mechanism in a useful way to heat or cool a home efficiently, such as using a fan

Operation

Creating forced convection is as easy as turning on a fan. Air is heated in the furnace and pushed through the house by the *blower*, which is a fan inside the ventilation system.

This blower outputs a specific quantity of air, and this output air flow is divided among all of the output grills (also called heater vents) in a home. Once it has traveled through the vents by being pushed through by fans, the warm, treated air is ejected through floor or ceiling vents into the rooms of a house. With help from natural convection this air then travels through the room, warming the room as it rises to the top by natural convection and slowly falls down to the floor as it cools. The system of heating the air and pushing it throughout the house to warm it then begins again.

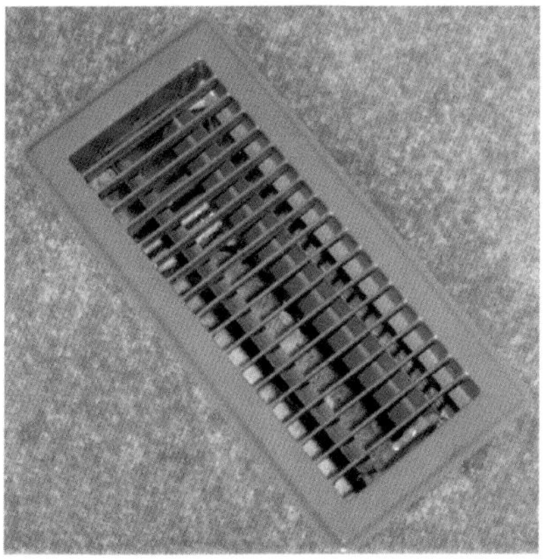

A floor heat register is part of the HVAC system that creates forced convection in a home

How the treated air gets to the output vents makes a difference, as the structure of the ductwork can create resistance to airflow at elbows, divisions, or places where the ductwork size changes. This change in turn effects how well this forced air system can heat a home since they are all sharing the output flow of air from one source —the furnace. Therefore, properly planning out the ducting is important. As a general rule, the best way for air to move through a duct is to have a straight duct, that is round in shape with a smooth inside wall—since curves and corners resist air flow. Wherever possible, this guideline should be followed to ensure that the air that is being forced out by the furnace heats the house properly. In addition, ensuring output vents are not covered by furniture or installed behind curtains ensures that the warm air output by the furnace is able to circulate throughout the room.

It is a common misconception that the more air that flows from a fan—or the more a fan "pushes" the air— the greater the effects of forced convection will occur, due to the large amount of heated or cooled air being pushed out by the fan. However, this is not entirely true. Part of how air moves through a home or other building has to do with the pressureand temperature that exists in the room before more air is pushed through. For example, if a room has a cold spot and the goal is to heat the room evenly, the pressure change in the area between the cold and warm spots, known as transitional "warm" area, factors into how well a fan will be able to move warm air to the cold area. If the pressure drop in this warm area is higher, there will be a lower flow rate of air into the cold section of the room, making it more difficult for the fan to push warm air into this section. This phenomenon is known as the *pressure drop* over the *heat sink* and it can be summed up easily by saying that it is more difficult for a fan to push warm or cool air through a region between two areas of different temperatures that also has a large pressure difference across its boundary.

Mechanism of Forced Convection

Convection heat transfer is complicated since it involves fluid motion as well as heat conduction. The fluid motion enhances heat transfer (the higher the velocity the higher the heat transfer rate).

The rate of convection heat transfer is expressed by Newton's law of cooling:

$$\dot{q}_{conv} = h(T_s - T_\infty) \qquad (W/m^2)$$

$$\dot{Q}_{conv} = hA(T_s - T_\infty) \qquad (W)$$

The convective heat transfer coefficient h strongly depends on the fluid properties and roughness of the solid surface, and the type of the fluid flow (laminar or turbulent).

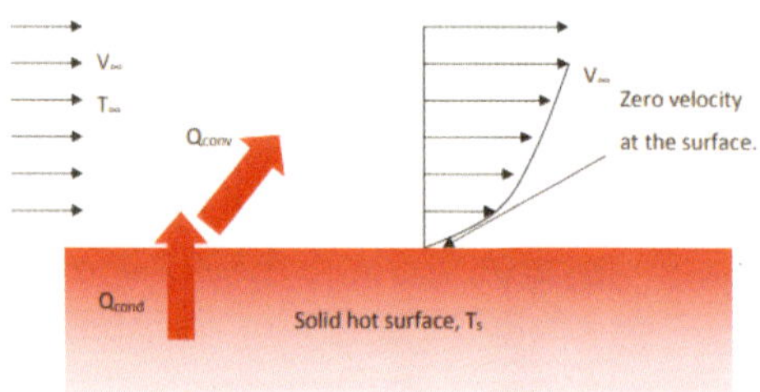

Forced convection

It is assumed that the velocity of the fluid is zero at the wall, this assumption is called no- slip condition. As a result, the heat transfer from the solid surface to the fluid layer adjacent to the surface is by pure conduction, since the fluid is motionless. Thus,

$$\left.\begin{aligned} \dot{q}_{conv} = \dot{q}_{cond} = -k_{fluid} \left.\frac{\partial T}{\partial y}\right|_{y=0} \\ \dot{q}_{conv} = h(T_s - T_\infty) \end{aligned}\right\} \rightarrow h = \frac{-k_{fluid} \left.\frac{\partial T}{\partial y}\right|_{y=0}}{T_s - T_\infty} \quad (W/m^2.K)$$

The convection heat transfer coefficient, in general, varies along the flow direction. The mean or average convection heat transfer coefficient for a surface is determined by (properly) averaging the local heat transfer coefficient over the entire surface.

Ceiling Fans

The use of ceiling fans in a home also represents a different type of forced convection. Ceiling fans can be used in both the winter and the summer, but their settings must be different in order to perform the desired task. In the summer months, the fan is generally set to a higher speed. The angle of the blades forces air down through the room. Generally this corresponds to a counter-clockwise rotation when looking at the fan from below. This downward breeze aids in the evaporation of perspiration from inhabitants of the home, cooling them. In the winter months, the fan must be used on a slower speed. The blades also spin in a different direction, generally clockwise when looking from below the fan, which pulls the cooler air up from the lower parts of the room. The cooler air from below then mixes with the warmer air that has risen and mixes the two, distributing warmer air throughout the building.

In the summer, ceiling fans should rotate counterclockwise to mix warm air and force a cool breeze downwards, creating a downdraft

In the winter, ceiling fans should rotate clockwise to pull cool air up from the room and force warm air downwards, creating an updraft

Mixed Convection

In any forced convection situation, some amount of natural convection is always present whenever there is g-forces present (i.e., unless the system is in free fall). When the natural convection is not negligible, such flows are typically referred to as mixed convection.

Mathematical Analysis

When analyzing potentially mixed convection, a parameter called the Archimedes number (Ar) parametrizes the relative strength of free and forced convection. The Archimedes number is the ratio of Grashof number and the square of Reynolds number, which represents the ratio of buoyancy force and inertia force, and which stands in for the contribution of natural convection. When Ar >> 1, natural convection dominates and when Ar << 1, forced convection dominates.

$$\mathrm{Ar} = \frac{\mathrm{Gr}}{\mathrm{Re}^2}$$

When natural convection isn't a significant factor, mathematical analysis with forced convection theories typically yields accurate results. The parameter of importance in forced convection is the Peclet number, which is the ratio of advection (movement by currents) and diffusion (movement from high to low concentrations) of heat.

$$\mathrm{Pe} = \frac{UL}{\alpha}$$

When the Peclet number is much greater than unity (1), advection dominates diffusion. Similarly, much smaller ratios indicate a higher rate of diffusion relative to advection.

Mixed Combined Convection

Mixed (combined) convection is a combination of forced and free convections which is the general case of convection when a flow is determined simultaneously by both an outer forcing system (i.e., outer energy supply to the fluid-streamlined body system) and inner volumetric (mass) forces, viz., by the nonuniform density distribution of a fluid medium in a gravity field. The most vivid manifestation of mixed convection is the motion of the temperature stratified mass of air and water areas of the Earth that the traditionally studied in geophysics. However, mixed convection is found in the systems of much smaller scales, i.e., in many engineering devices. We shall illustrate this on the basis of some examples referring to channel flows, the most typical and common cases. On heating or cooling of channel walls, and at the small velocities of a fluid flow that are characteristic of a laminar flow, mixed convection is almost always realized. Pure forced laminar convection may be obtained only in capillaries. Studies of turbulent channel flows with substantial gravity field effects have actively developed since the 1960s after their becoming important in engineering practice by virtue of the growth of heat loads and channel dimensions in modern technological applications (thermal and nuclear power engineering, pipeline transport).

In a mathematical description of mixed convection in the equations of motion (the Navier-Stokes equations), both the term characterizing the pressure head loss dp/dx and the term characterizing mass forces ρ_g are retained. The simplest case allowing an analytical solution of the problem refers to the steady-state laminar flow at a distance from the tube inlet with a constant heat flux on a wall (q_w = const). Figure below shows the calculated velocity and temperature distributions both in an ascending flow, i.e., when the directions of forced and free near-wall convections coincide (case "a"), and in a descending flow in heated tubes, i.e., when the directions of forced and free near-wall convections are opposite (case "b"). The curves refer to different values of the Rayleigh number $Ra_A = g\beta D^4 (dT/dx)/\nu^2 = 4\ Gr_q/Re = (4/Re)(g\beta q_w D^4/\lambda\nu^2)$. In the case of an ascending fluid flow, velocity profiles peak near the wall giving an M-like shape which becomes sharper with the growth of the effect of thermogravitation convection until instability of the laminar flow occurs before the calculated values of the axis velocity have vanished (figure below a, Ra_A = 625). In a descending flow the stability is disturbed rather quickly, as indicated by the velocity profiles (figure below b) which even at relatively small growth of the bouyancy effect acquire zero values of the velocity gradient on the wall. A remarkable property of temperature distributions should be noted for all the cases considered (figure below a and b); even with strong deformation of velocity distribution, the temperature profiles only differ slightly from the profiles for pure forced convection. The shown specific features of velocity and temperature shown in figure below are reflected on the corresponding variation of the relative heat transfer value presented in the left portion of figure on the next pages where Nu_o means the Nusselt number for a "purely" forced flow.

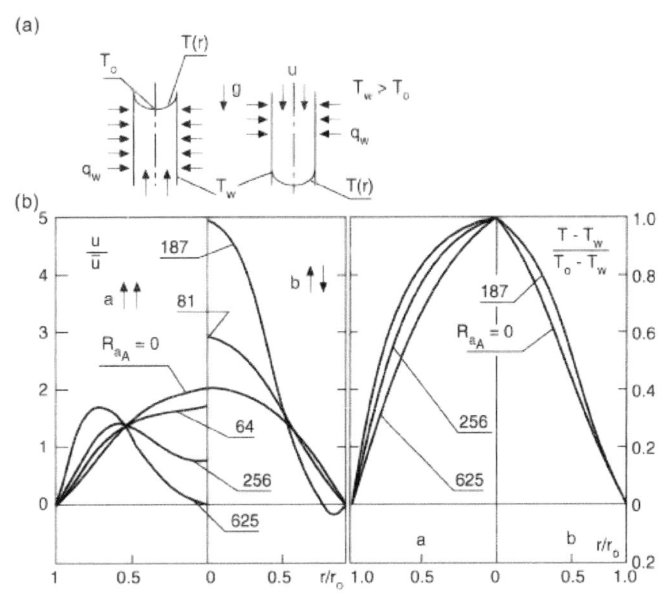

Velocity and temperature profiles in mixed convection

The variation of heat transfer with Grq in turbulent and laminar mixed convection is considerably different figure below. In the case of a descending flow (b) (Nu/Nu_0), falls with Gr_q in laminar mixed convection due to the retardation near a wall, but with turbulent mixed convection (Nu/Nu_0) grows as the heat load increases due to additional flow turbulization. For an ascending turbulent flow the behavior of the curve (Nu/Nu_0) — Gr_q is nonmonotonic. First, heat transfer deteriorates with increasing Gr_q. This is caused by the attenuation of turbulent momentum and heat transfer as a result of the effect of buoyancy forces on the shear stress profile and hence as the generation of turbulence. With further increase in the Grashof number, heat transfer begins to grow, thus reflecting an intensive development of the free convection effect on the flow as a whole as also takes place in the laminar mode (case "a"). In turbulent flow, the velocity profiles also acquire a typical M-like shape, but the quantitative characteristics, as well as the mechanism of momentum and heat transfer, are quite different for turbulent and laminar mixed convection. In the region of a strong effect of free convection the mode of thermal turbulent convection with the dependence of the Nusselt number on the Grashof number, characteristic to turbulent free convection $Nu = A(Pr)Gr^{1/4}$ and excluding the effect of a geometry parameter on heat transfer, is established. Line B in figure below is constructed by the relation for the Nusselt number with developed free convection on a vertical plate. Under normal conditions (in particular, for air at atmospheric pressure and temperatures close to room ones, the data of which are given in figure below) the effect of the gravity field on a forced turbulent flow manifests itself only at relatively small Reynolds numbers, of the order of 10^4 for tubes, and with channels of rather large dimensions equal with characteristic equivalent diameters of the order of 1 m and more. However, for media with strong density variation, e.g., under near-critical conditions where density varies from the values typical of gases to the values characteristic for a liquid, mixed convection is realized in tubes of small diameters (of the order of several millimeters) and at large values of the Reynolds number.

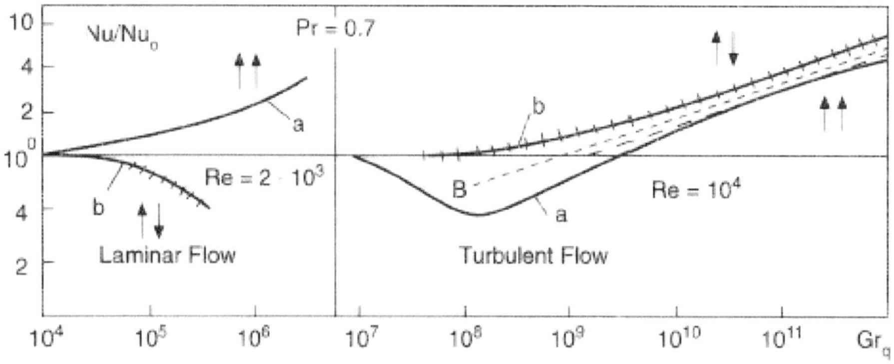

Variation of Nusselt number Nu (= $\alpha D/\lambda$, where α = least transfer coefficient, D = tube diameter and λ = thermal conductivity) with Gr_q (= $g\beta\, \dot{q}_w\, D^4/\lambda w$)

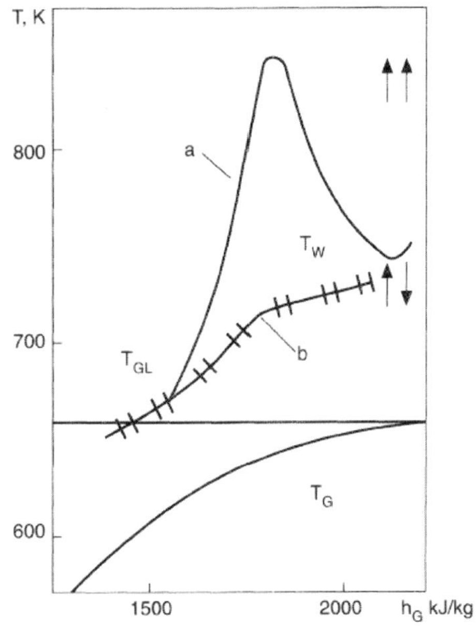

Temperature profiles as a function of fluid enthalphy (h_{GC}) for ascending (a) and descending (b) flow in a 3 mm tube at (p/pc) = 1.1. (T_w = wall temperature, G_{ct} = mean fluid temperature)

In a turbulent flow in the presence of heat transfer, the gravity field leads not only to the manifestation of large-scale free convection covering the entire flow as in the laminar mode, but also to the manifestation of local effects on turbulence that radically change the character of heat transfer. One may judge the character of the manifestation of the effect of the buoyancy forces on turbulence from the velocity distributions given in figure below. These velocity profiles are plotted in the near-wall region in terms of the universal parameters $u^+ = u/u^*$ and $y^- = u^*\rho y/\eta$, where u^* is the friction velocity ($\sqrt{\tau_w/\rho}$) and y^+ the distance from the wall. The data are for a horizontal channel. The channel can either be heated from above giving stable stratification or from below giving unstable stratification. Points 1 and 2 represent the profile near the bottom wall for a case with stable density distribution (stable stratification) and from points 4 and 5 show the profile with unstable density distribution (unstable stratification); those results may be compared with the universal velocity distribution in an equilibrium wall flow (points 3). Points 2 and 4 indicate to a small effect of density stratification on turbulence. With strong stable stratification, turbulent transport of fluid particles is retarded and a turbulent flow, which occurred under isothermal conditions, laminarizes at the same Reynolds number and the velocity distribution in this case (points 1) approaches the distribution typical of a laminar flow (a dashed line). With a strong unstable stratification (points 5) the buoyancy forces additionally turbulize the flow and in the limited case the wall logarithmic law for a velocity is transformed to the "−1/3"-law, viz.: $u \sim y^{-3}$ (line 6) found when studying the near-Earth atmospheric turbulence. The data indicate that in physical studies and mathematical simulation of mixed convection, one should simultaneously take into account both the global effect of gravitation on the flow as a whole and its direct effect on turbulence.

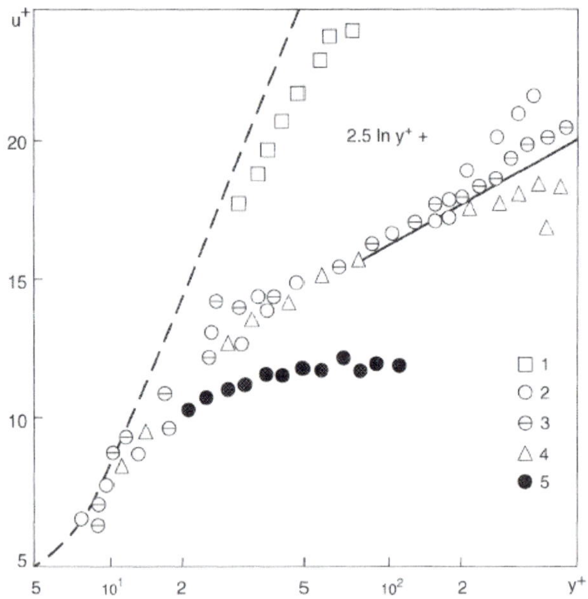

Dimensionless velocity profiles in mixed convection

References

- Incropera, Frank P.; DeWitt, David P. (1990). Fundamentals of Heat and Mass Transfer (3rd ed.). John Wiley & Sons. p. 28. ISBN 0-471-51729-1
- Raats, P. A. C. (1969). "Steady Gravitational Convection Induced by a Line Source of Salt in a Soil". Soil Science Society of AmericaProceedings. 33 (4): 483. Bibcode:1969SSASJ..33..483R. doi:10.2136/sssaj1969.03615995003300040005x
- Pidwirny, Michael (2008). "CHAPTER 8: Introduction to the Hydrosphere (e). Cloud Formation Processes". Physical Geography. Archived from the original on 2008-12-20. Retrieved 2009-01-01
- Hahn, Douglas G.; Manabe, Syukuro (1975). "The Role of Mountains in the South Asian Monsoon Circulation". Journal of Atmospheric Sciences. 32 (8): 1515–1541. Bibcode:1975JAtS...32.1515H. doi:10.1175/1520-0469(1975)032<1515:TROMIT>2.0.CO;2. ISSN 1520-0469
- Matthias Tomczak and J. Stuart Godfrey (2001). Regional Oceanography: an Introduction. Archived 2009-09-14 at the Wayback Machine. Matthias Tomczak, pp. 42. ISBN 81-7035-306-8
- McKinnon, William B. (2006). "On convection in ice I shells of outer Solar System bodies, with detailed application to Callisto". Icarus. 183: 435–450. Bibcode:2006Icar..183..435M. doi:10.1016/j.icarus.2006.03.004
- "Fog And Stratus – Meteorological Physical Background". Zentralanstalt für Meteorologie und Geodynamik. FMI. 2007. Archived from the original on 2011-07-06. Retrieved 2009-02-07

4

Heat Exchanger and its Types

Heat can be transferred between two or more fluids using a device called a heat exchanger. It can be of different types such as regenerative heat exchanger, plate and frame heat exchanger, shell and tube heat exchanger and dynamic scraped surface heat exchanger, among others. This chapter has been carefully written to provide an easy understanding of heat exchanger and its varied types.

Heat Exchanger

A heat exchanger is a device used to transfer heat between two or more fluids. The fluids can be single or two phase and, depending on the exchanger type, may be separated or in direct contact. Devices involving energy sources such as nuclear fuel pins or fired heaters are not normally regarded as heat exchangers although many of the principles involved in their design are the same.

In order to discuss heat exchangers it is necessary to provide some form of categorization. There are two approaches that are normally taken. The first considers the flow configuration within the heat exchanger, while the second is based on the classification of equipment type primarily by construction. Both are considered here.

Classification of Heat Exchangers by Flow Configuration

There are four basic flow configurations:

- Counter Flow
- Cocurrent Flow
- Crossflow
- Hybrids such as Cross Counterflow and Multi Pass Flow

Figure below illustrates an idealized counterflow exchanger in which the two fluids flow parallel to each other but in opposite directions. This type of flow arrangement allows the largest change in temperature of both fluids and is therefore most efficient (where efficiency is the amount of actual heat transferred compared with the theoretical maximum amount of heat that can be transferred).

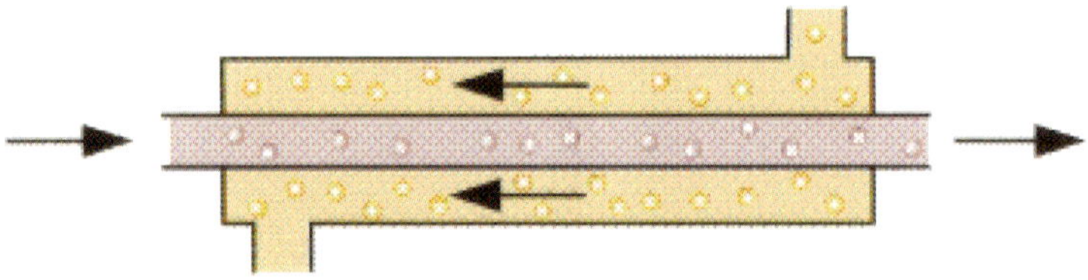

Countercurrent flow

In cocurrent flow heat exchangers, the streams flow parallel to each other and in the same direction as shown in

Figure below, This is less efficient than countercurrent flow but does provide more uniform wall temperatures.

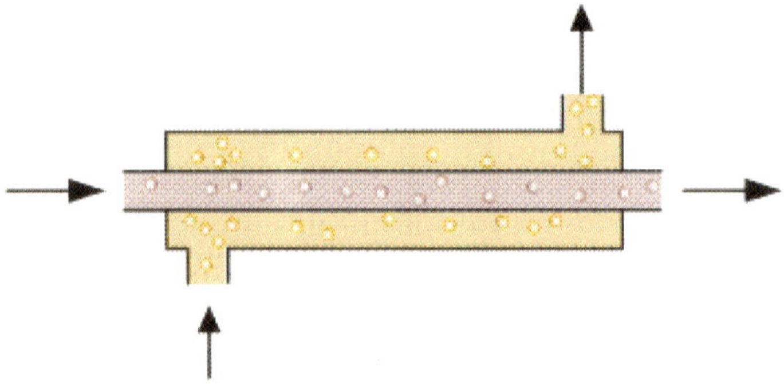

Cocurrent flow

Crossflow heat exchangers are intermediate in efficiency between countercurrent flow and parallel flow exchangers. In these units, the streams flow at right angles to each other as shown in figure below.

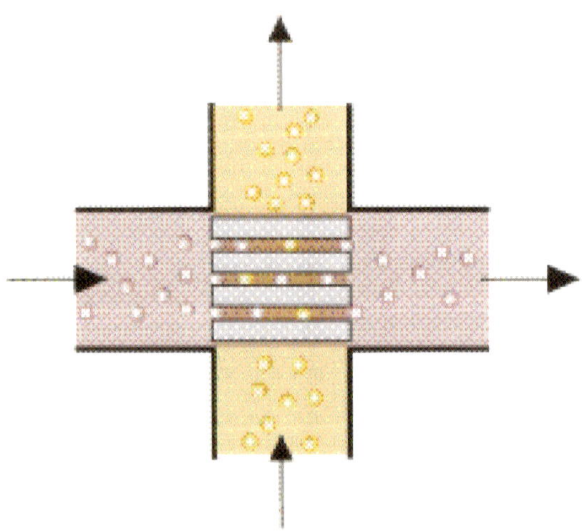

Crossflow

In industrial heat exchangers, hybrids of the above flow types are often found. Examples of these are combined crossflow/counterflow heat exchangers and multi pass flow heat exchangers.

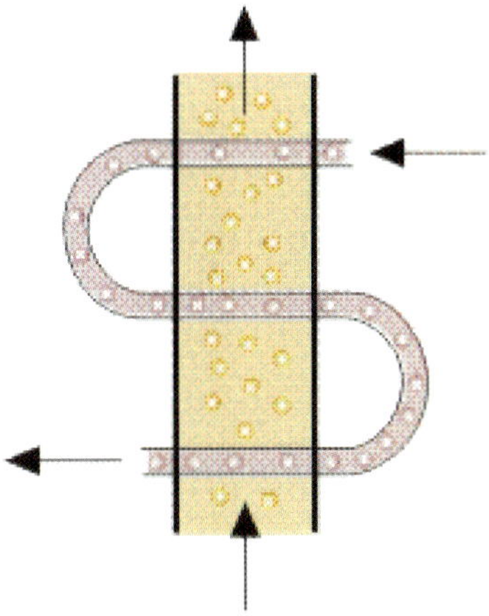

Cross/counter flow

Classification of Heat Exchangers by Construction

Heat exchangers are classified mainly by their construction, Garland (1990). The first level of classification is to divide heat exchanger types into recuperative or regenerative. A *Recuperative Heat Exchanger* has separate flow paths for each fluid and fluids flow simultaneously through the exchanger exchanging heat across the wall separating the flow paths. A Regenerative Heat Exchanger has a single flow path, which the hot and cold fluids alternately pass through.

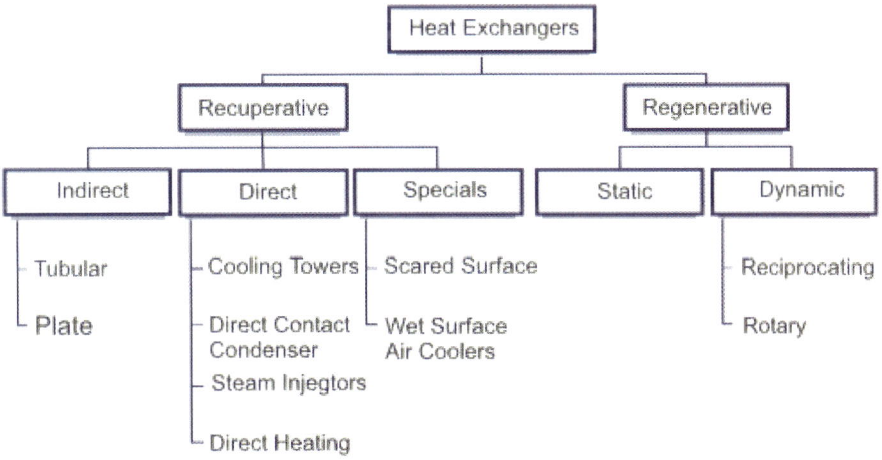

Heat exchanger classifications

Regenerative Heat Exchangers

In a regenerative heat exchanger, the flow path normally consists of a matrix, which is heated when the hot fluid passes through it (this is known as the "hot blow"). This heat is then released to the cold fluid when this flows through the matrix (the "cold blow"). Regenerative Heat Exchangers are sometimes known as *Capacitive Heat Exchangers*. A good overview of regenerators is provided by Walker (1982).

Regenerators are mainly used in gas/gas heat recovery applications in power stations and other energy intensive industries. The two main types of regenerator are Static and Dynamic. Both types of regenerator are transient in operation and unless great care is taken in their design there is normally cross contamination of the hot and cold streams. However, the use of regenerators is likely to increase in the future as attempts are made to improve energy efficiency and recover more low grade heat. However, because regenerative heat exchangers tend to be used for specialist applications recuperative heat exchangers are more common.

Recuperative Heat Exchangers

There are many types of recuperative exchangers, which can broadly be grouped into indirect contact, direct contact and specials. Indirect contact heat exchangers keep the fluids exchanging heat separate by the use of tubes or plates etc.. Direct contact exchangers do not separate the fluids exchanging heat and in fact rely on the fluids being in close contact.

Heat Exchanger Types

This section briefly describes some of the more common types of heat exchanger and is arranged according to the classification given in figure above.

Indirect Heat Exchangers

In this type, the steams are separated by a wall, usually metal. Examples of these are tubular exchangers.

Tubular heat exchangers are very popular due to the flexibility the designer has to allow for a wide range of pressures and temperatures. Tubular heat exchangers can be subdivided into a number of categories, of which the shell and tube exchanger is the most common.

A Shell and Tube Exchanger consists of a number of tubes mounted inside a cylindrical shell. Figure given below illustrates a typical unit that may be found in a petrochemical plant. Two fluids can exchange heat, one fluid flows over the outside of the tubes while the second fluid flows through the tubes. The fluids can be single or two phase and can flow in a parallel or a cross/counter flow arrangement. The shell and tube exchanger consists of four major parts:

- Front end–this is where the fluid enters the tubeside of the exchanger.
- Rear end–this is where the tubeside fluid leaves the exchanger or where it is returned to the front header in exchangers with multiple tubeside passes.
- Tube bundle–this comprises of the tubes, tube sheets, baffles and tie rods etc. to hold the bundle together.
- Shell—this contains the tube bundle.

The popularity of shell and tube exchangers has resulted in a standard being developed for their designation and use. This is the Tubular Exchanger Manufactures Association (TEMA) Standard. In general shell and tube exchangers are made of metal but for specialist applications (e.g., involving strong acids of pharmaceuticals) other materials such as graphite, plastic and glass may be used. It is also normal for the tubes to be straight but in some cryogenic applications helical or *Hampson coils* are used. A simple form of the shell and tube exchanger is the Double Pipe Exchanger. This exchanger consists of a one or more tubes contained within a larger pipe. In its most complex form there is little difference between a multi tube double pipe and a shell and tube exchanger. However, double pipe exchangers tend to be modular in construction and so several units can be bolted together to achieve the required duty.

Other types of tubular exchanger include:

- Furnaces—the process fluid passes through the furnace in straight or helically wound tubes and the heating is either by burners or electric heaters.
- Tubes in plate—these are mainly found in heat recovery and air conditioning applications. The tubes are normally mounted in some form of duct and the plates act as supports and provide extra surface area in the form of fins.
- Electrically heated–in this case the fluid normally flows over the outside of electrically heated tubes.
- Air Cooled Heat Exchangers consist of bundle of tubes, a fan system and supporting structure. The tubes can have various type of fins in order to provide additional surface area on the air side. Air is either sucked up through the tubes by a fan mounted above the bundle (induced draught) or blown through the tubes by a fan mounted under the bundle (forced draught). They tend to be used in locations where there are problems in obtaining an adequate supply of cooling water.
- Heat Pipes, Agitated Vessels and Graphite Block Exchangers can be regarded as tubular or could be placed under Recuperative "Specials". A heat pipe consists of a pipe, a wick material and a working fluid. The working fluid absorbs heat, evaporates and passes to the other end of the heat pipe were it condenses and releases heat. The fluid then returns by capillary action to the hot end of the heat pipe to re-evaporate. Agitated vessels are mainly used to heat viscous fluids. They consist of a vessel with tubes on the inside and an agitator such as a propeller or a helical ribbon impeller. The tubes carry the hot fluid and the agitator is introduced to ensure uniform heating of the cold fluid. Carbon block exchangers are normally used when corrosive fluids need to be heated or cooled. They consist of solid blocks of carbon which have holes drilled in them for the fluids to pass through. The blocks are then bolted together with headers to form the heat exchanger.

Plate heat exchangers separate the fluids exchanging heat by the means of plates. These normally have enhanced surfaces such as fins or embossing and are either bolted together, brazed or welded. Plate heat exchangers are mainly found in the cryogenic and food processing industries. However, because of their high surface area to volume ratio, low inventory of fluids and their ability to handle more than two steams, they are also starting to be used in the chemical industry.

Plate and Frame Heat Exchangers consist of two rectangular end members which hold together a number of embossed rectangular plates with holes on the corner for the fluids to pass through. Each of the plates is separated by a gasket which seals the plates and arranges the flow of fluids between the plates, figure below. This type of exchanger is widely used in the food industry because it can easily be taken apart to clean. If leakage to the environment is a concern it is possible to weld two plate together to ensure that the fluid flowing between the welded plates can not leak. However, as there are still some gaskets present it is still possible for leakage to occur. Brazed plate heat exchangers avoid the possibility of leakage by brazing all the plates together and then welding on the inlet and outlet ports.

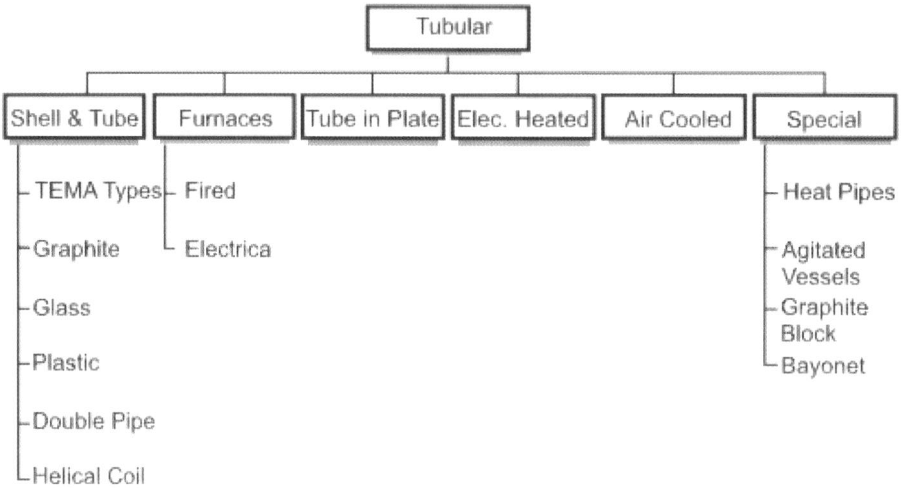

Tubular exchanger classification

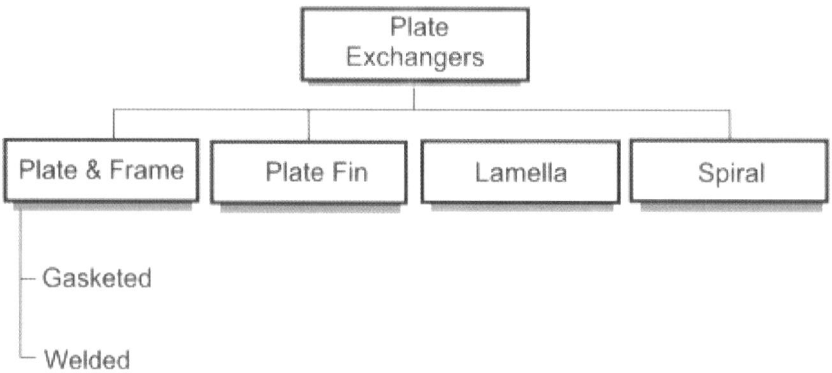

Plate exchanger classification

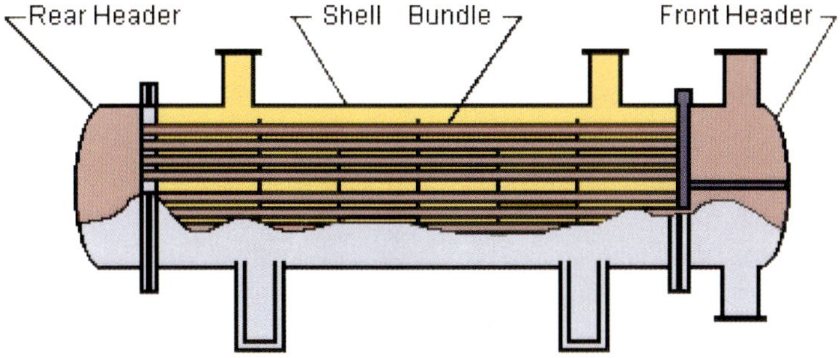

Shell and tube exchanger

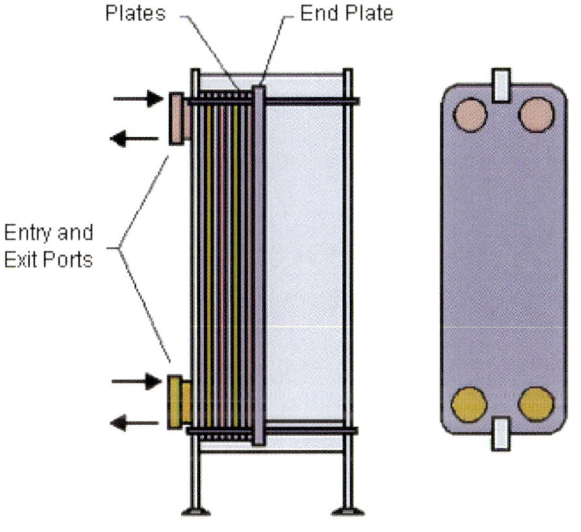

Plate and frame exchanger

Plate Fin Exchangers consist of fins or spacers sandwiched between parallel plates. The fins can be arranged so as to allow any combination of crossflow or parallel flow between adjacent plates. It is also possible to pass up to 12 fluid streams through a single exchanger by careful arrangement of headers. They are normally made of aluminum or stainless steel and brazed together. Their main use is in gas liquefaction due to their ability to operate with close temperature approaches.

Lamella heat exchangers are similar in some respects to a shell and tube. Rectangular tubes with rounded corners are stacked close together to form a bundle, which is placed inside a shell. One fluid passes through the tubes while the fluid flows in parallel through the gaps between the tubes. They tend to be used in the pulp and paper industry where larger flow passages are required.

Spiral plate exchangers are formed by winding two flat parallel plates together to form a coil. The ends are then sealed with gaskets or are welded. They are mainly used with viscous, heavily fouling fluids or fluids containing particles or fibres.

Direct Contact

This category of heat exchanger does not use a heat transfer surface, because of this, it is often cheaper than indirect heat exchangers. However, to use a direct contact heat exchanger with two fluids they must be immiscible or if a single fluid is to be used it must undergo a phase change.

The most easily recognizable form of direct contact heat exchanger is the natural draught Cooling Tower found at many power stations. These units comprise of a large approximately cylindrical shell (usually over 100 m in height) and packing at the bottom to increase surface area. The water to be cooled is sprayed onto the packing from above while air flows in through the bottom of the packing and up through the tower by natural buoyancy. The main problem with this and other types of direct contact cooling tower is the continuous need to make up the cooling water supply due to evaporation.

Direct contact condensers are sometimes used instead of tubular condensers because of their low capital and maintenance costs. There are many variations of direct contact condenser. In its simplest form a coolant is sprayed from the top of a vessel over vapor entering at the side of the vessel. The condensate and coolant are then collected at the bottom. The high surface area achieved by the spray ensures they are quite efficient heat exchangers.

Steam injection is used for heating fluids in tanks or in pipelines. The steam promotes heat transfer by the turbulence created by injection and transfers heat by condensing. Normally no attempt is made to collect the condensate.

Direct heating is mainly used in dryers where a wet solid is dried by passing it through a hot air stream. Another form of direct heating is Submerged Combustion. This was developed mainly for the concentration and crystallization of corrosive solutions. The fluid is evaporated by the flame and exhaust gases being aimed down into the fluid which is held in some form of tank.

Specials

The wet surface air cooler is similar in some respects to an air cooled heat exchanger. However, in this type of unit water is sprayed over the tubes and a fan sucks air and the water down over the tube bundle. The whole system is enclosed and the warm damp air is normally vented to atmosphere.

Scraped Surface Exchangers consist of a jacketed vessel which the fluid passes through and a rotating scraper which continuously removes deposit from the inside walls of the vessel. These units are used in the food and pharmaceutical industry in process where deposits form on the heated walls of the jacketed vessel.

Static Regenerators

Static regenerators or fixed bed regenerators have no moving parts except for valves. In this case the hot gas passes through the matrix for a fixed time period at the end of which a reversal occurs, the hot gas is shut off and the cold gas passes through the matrix. The main problem with this type of unit is that both the hot and cold flow are intermittent. To overcome this and have continuous operation at least two static regenerators are required or a rotary regenerator could be used.

Rotary Regenerator

In a rotary regenerator cylindrical shaped packing rotates about the axis of a cylinder between a pair of gas seals. Hot and cold gas flows simultaneously through ducting on either side of the gas seals and through the rotating packing.

Thermal Analysis

The thermal analysis of any heat exchanger involves the solution of the basic heat transfer equation.

$$d\dot{Q} = \alpha(T_h - T_c)dA$$

This equation calculates the amount of heat $d\dot{Q}$ transferred through the area dA, where T_h and T_c are the local temperatures of the hot and cold fluids, α is the local heat transfer coefficient and dA is the local incremental area on which α is based. For a flat wall

$$\alpha_w = \delta_w / \lambda_w$$

where δ_w is the wall thickness and λ_w its thermal conductivity.

For single phase flow past the wall α for each of the streams is a function of Re and Pr. When condensing or boiling is taking place α may also be a function of the temperature difference. Once the heat transfer coefficient for each stream and the wall are known the overall heat transfer coefficient U is then given by

$$1/U = 1/\alpha_h + r_w + 1/\alpha_c$$

where the wall resistance r_w is given by $1/\alpha_w$. The total rate of heat transfer between the hot and cold fluids is then given by

$$\dot{Q}_T = UA_T(T_h - T_c)$$

This equation is for constant temperatures and heat transfer coefficients. In most heat exchangers this is not the case and so a different form of the equation is used

$$\dot{Q}_T = UA_T \Delta T_M$$

where $\dot{Q}_T$ is the total heat load, U is the mean overall heat transfer coefficient and ΔT_M the mean temperature difference.

The calculation of ΔT_M and the removal of the constant heat transfer coefficient assumption is described in:

Mean Temperature Difference

Calculation of U and ΔT_M requires information on the exchanger type, the geometry (e.g., the size of the passages in a plate or the diameter of a tube), flow orientation, pure countercurrent flow or crossflow, etc. The total duty $\dot{Q}_T$ can then be calculated using an assumed value of AT and compared with the required duty. Changes to the assumed geometry can then be made and U, ΔT_M and $\dot{Q}_T$ recalculated to eventually iterate to a solution where $\dot{Q}_T$ is equal to the required duty. However, in performing the thermal analysis a check should also be made at each iteration that the allowable pressure drop is not exceeded. Computer programs such as TASC from HTFS (Heat Transfer and Fluid Flow Service) perform these calculations automatically and optimize the design.

Mechanical Considerations

All heat exchangers types have to undergo some form of mechanical design. Any exchanger that operates at above atmospheric pressure should be designed according to the locally specified *pressure vessel design code* such as ASME VIII (American Society of Mechanical Engineers) or BS 5500 (British Standard). These codes specify the requirements for a pressure vessel, but they do not deal with any specific features of a particular heat exchanger type. In some cases specialist standards exist for certain types of heat exchanger. Two of these are listed below, but in general individual manufacturers define their own standards.

A heat pump extracts heat from a natural geothermal hot spring, used to heat buildings at Hot Springs Lodge and Pool in Glenwood Springs, Colorado. The exchanger is the algae-covered plate full of copper tubes in the center of the water

You can see heat exchangers in all kinds of places, usually working to heat or cool buildings or helping engines and machines to work more efficiently. Refrigerators and air-conditioners, for example, use heat exchangers in the opposite way from central heating systems: they remove heat from a compartment or room where it's not wanted and pump it away in a fluid to some other place where it can be dumped out of the way.

In power plants or engines, exhaust gases often contain heat that's heading uselessly away into the open air. That's a waste of energy and something a heat exchanger can certainly reduce (though not eliminate entirely—some heat is always going to be lost). The way to solve this problem is with heat exchangers positioned inside the exhaust tail pipes or smokestacks. As the hot exhaust gases drift upward, they brush past copper fins with water flowing through them. The water carries the heat away, back into the plant. There, it might be recycled directly, maybe warming the cold gases that feed into the engine or furnace, saving the energy that would otherwise be needed to heat them up. Or it could be put to some other good use, for example, heating an office near the smokestack.

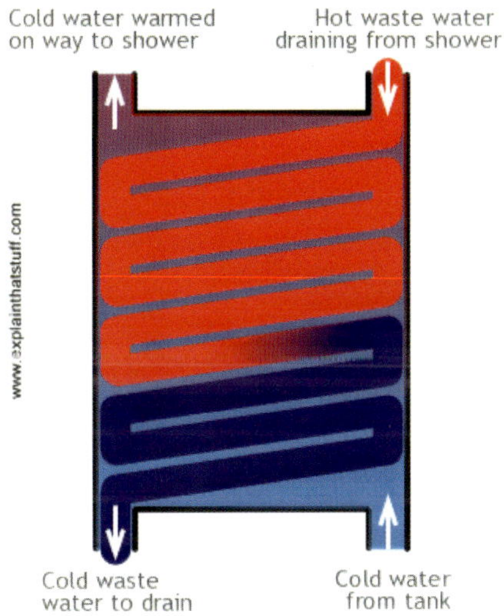

How a shower waste-water heat exchanger works. Hot outgoing waste-water warms incoming cold water, reducing the energy you need to get the water hot and making the whole thing more efficient

In buses, fluid used to cool down the diesel engine is often passed through a heat exchanger and the heat it reclaims is used to warm cold air from outside that is pumped up from the floor of the passenger compartment. That saves the need for having additional, wasteful electric heaters inside the bus. A car radiator is another kind of heat exchanger. Water that cools the engine flows through the radiator, which has lots of parallel, aluminum fins open to the air. As the car drives along, cold air blowing past the radiator removes some of the heat, cooling the water and heating the air and keeping the engine working efficiently. The radiator›s waste heat is used to heat the passenger compartment, just like on a bus.

If you have an energy-efficient shower, it might have a heat exchanger installed in the wastewater outlet. As the water drips past your body and down the plug, it runs through the copper coils of a heat exchanger. Meanwhile, cold water that's feeding into the shower to be heated pumps up past the same coils, not mixing with the dirty water but picking up some of its waste heat and warming slightly—so the shower doesn't need to heat it so much.

Condenser

The change from liquid phase to vapor phase is called vaporization and the reverse phase transfer is condensation. The change from liquid to vapor or vapor to liquid occurs at one temperature (called saturation or equilibrium temperature) for a pure fluid compound at a given pressure. The industrial practice of vaporization and condensation occurs at almost constant pressure; therefore the phase change occurs isothermally.

Condensation occurs by two different physical mechanisms i.e. drop-wise condensation and film condensation.

The nature of the condensation depends upon whether the condensate (liquid formed from vapor) wets or does not wet the solid surface. If the condensate wets the surface and flows on the surface in the form of a film, it is called film condensation. When the condensate does not wet the solid surface and the condensate is accumulated in the form of droplets, is drop-wise condensation. Heat transfer coefficient is about 4 to 8 times higher for drop wise condensation. The condensate forms a liquid film on the baresurface in case of film condensation. The heat transfer coefficient is lower for film condensation due to the resistance of this liquid film.

Dropwise condensation occurs usually on new, clean and polished surfaces. The heat exchanger used for condensation is called condenser. In industrial condensers, film condensation normally occurs.

Types of Condensers

There two general types of condensers:

i. Vertical condenser

   Downflow vertical condenser: The vapor enters at the top of condenser and flows down inside tubes. The condensate drains from the tubes by gravity and vapor induced shear first figure below.

   Upflow vertical condenser: In case of upflow condenser, the vapor enters at the bottom and flows upwards inside the tubes. The condensate drains down the tubes by gravity only.

ii. Horizontal condenser: The condensation may occur inside or outside the horizontal tubes second figure below. Condensation in the tube-side is common in aircooled condensers. The main disadvantage of this type of condenser is that the liquid tends to build up in the tubes. Therefore the effective heat transfer coefficient is reduced significantly.

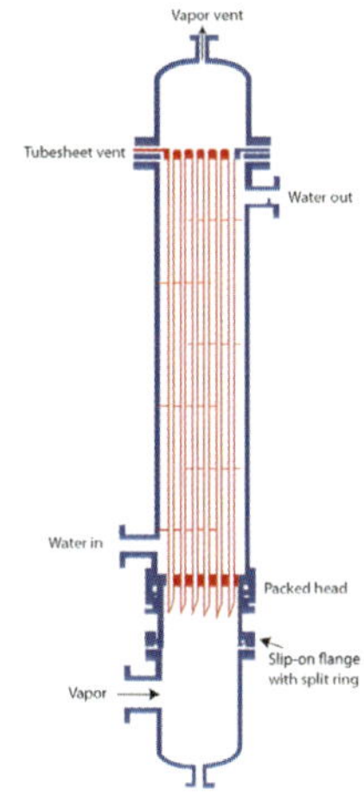

Downflow vertical condenser with condensation inside tube

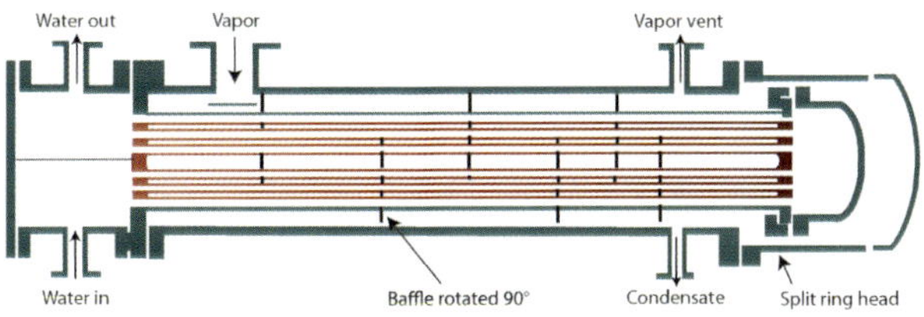

Horizontal condenser with condensation outside horizontal tubes

Condenser Design

The design of condenser is similar to a typical shell and tube exchangers. A condenser must have a vent for removal of non-condensable gas. The non-condensable gas decreases the heat transfer rate. Condenser usually use a wider baffle spacing of $B = D_s$ (ID of shell) as the allowable pressure drop in shell side vapor is usually less. Vertical cut-segmental baffles are generally used in condensers for side-to-side vapor flow and not for top to bottom. An opening at the bottom of the baffles is provided to allow draining of condensate.

Mean Temperature Difference

The condensation occurs almost at a fixed temperature (isothermally) at constant pressure for a pure saturated vapor compound. The logarithmic mean temperature difference can be used for condenser design. No correction factor for multiple pass condensers is required. The logarithmic mean temperature difference:

$$\text{LMTD} = \frac{(T_{sat} - t_1) - (T_{sat} - t_2)}{\ln \dfrac{(T_{sat} - t_1)}{(T_{sat} - t_2)}} = \frac{(t_2 - t_1)}{\ln \dfrac{(T_{sat} - t_1)}{(T_{sat} - t_2)}}$$

Where, T_{sat} = Saturation vapor temperature

t_1 = Coolant inlet temperature

t_2 = Coolant outlet temperature

Calculation of Heat Transfer Co-efficient During Condensation

Calculation of tube side heat transfer co-efficient (h_i): The calculation of heat transfer co-efficient for the cold fluid (coolant) can be performed similarly as discussed in design of shell and tube heat exchanger (heat transfer without phase change). Here it is assumed that the coolant flows the in tube side and the condensing saturated vapor flows in the shell side.

Calculation of shell-side heat transfer coefficient (condensing film heat transfer coefficient) (h_o): The Kern method is discussed here to calculate the individual heat transfer co-efficient of the condensing fluid by trial and error calculation.

i. Assume, $h_{0(assm)}$ in the range from 100 to 300 $BTU.h^{-1}.ft^{-2}.^{0}F^{-1}$. The film coefficient of condensing hydrocarbons generally varies in this range. Air-free condensing steam has a coefficient of 1500 $BTU.h^{-1}.ft^{-2}.^{0}F^{-1}$..

ii. Calculate the tube wall temperature (T_m):

$$T_w = T_{C(avg)} + \frac{h_0\left(T_v - T_{C(avg)}\right)}{\left(h_{i0} + h_0\right)}$$

Or

$$T_w = T_{CC} + \frac{h_0\left(T_v - T_{CC}\right)}{\left(h_{i0} + h_0\right)}$$

Where, $h_{io} = h_i \times \frac{d_i}{d_{i0}}$ (d_i tube ID and d_0 tube OD)

$T_{C(avg)}$ = Average temperature of the cold fluid

T_{CC} = Caloric temperature of the cold fluid

iii. Calculate condensate film temperature, $T_f = \frac{(T_w + T_V)}{2}$

T_V = Condensation temperature (For pure fluid compound T_V is the saturation temperature. Average of condensation over a temperature range also can be used for non-isothermal condensation).

iv. Calculate all thermophysical property of the condensing fluid at film temperature T_f.

v. Recalculate, $h_{0(cal)}$ from j_H factor.

Now again set, $h_{0(assm)} = h_{0(cal)}$ and continue the calculation till $h_{0(assm)} \approx h_{0(cal)}$

vi. Calculate the overall heat transfer-coefficient (U_d) including the dirt factors.

Examples of Condensers

- A surface condenser is one in which condensing medium and vapors are physically separated and used when direct contact is not desired. It is a shell and tube heat exchanger installed at the outlet of every steam turbine in thermal power stations. Commonly, the cooling water flows through the tube side and the steam enters the shell side where the condensation occurs on the outside of the heat transfer tubes. The condensate drips down and collects at the bottom, often in a built-in pan called a *hotwell*. The shell side often operates at a vacuum or partial vacuum, produced by the difference in specific volume between the steam and condensate. Conversely, the vapor can be fed through the tubes with the coolant water or air flowing around the outside.

- In chemistry, a condenser is the apparatus which cools hot vapors, causing them to condense into a liquid. See “Condenser (laboratory)” for laboratory-scale condensers, as opposed to industrial-scale condensers. Examples include the Liebig condenser, Graham condenser, and Allihn condenser. This is not to be confused with a condensation reaction which links two fragments into a single molecule by an addition reaction and an elimination reaction.

  In laboratory distillation, reflux, and rotary evaporators, several types of condensers are commonly used. The Liebig condenser is simply a straight tube within a cooling water jacket, and is the simplest (and relatively least expensive) form of condenser. The Graham condenser is a spiral tube within a water jacket, and the Allihn condenser has a series of large and small constrictions on the inside tube, each increasing the surface area upon which the vapor constituents may condense. Being more complex shapes to manufacture, these latter types are also more expensive to purchase. These three types of condensers are laboratory glasswareitems since they are typically made of glass. Commercially available condensers usually are fitted with ground glass joints and come in standard lengths of 100, 200, and 400 mm. Air-cooled condensers are unjacketed, while water-cooled condensers contain a jacket for the water.

- Larger condensers are also used in industrial-scale distillation processes to cool distilled vapor into liquid distillate. Commonly, the coolant flows through the tube side and distilled vapor through the shell side with distillate collecting at or flowing out the bottom.

Condenser unit for central air conditioning for a typical house

- A condenser unit used in central air conditioning systems typically has a heat exchanger section to cool down and condense incoming refrigerant vapor into liquid, a compressor to raise the pressure of the refrigerant and move it along, and a fan for blowing outside air through the heat exchanger section to cool the refrigerant inside. A typical configuration of such a condenser unit is as follows: The heat exchanger section wraps around the sides of the unit with the compressor inside. In this heat exchanger section, the refrigerant goes through multiple tube passes, which are surrounded by heat transfer fins through which cooling air can move from outside to inside the unit. There is a motorized fan inside the condenser unit near the top, which is covered by some grating to keep any objects from accidentally falling inside on the fan. The fan is used to blow the outside cooling air in through the heat exchange section at the sides and out the top through the grating. These condenser units are located on the outside of the building they are trying to cool, with tubing between the unit and building, one for vapor refrigerant entering and another for liquid refrigerant leaving the unit. Of course, an electric power supply is needed for the compressor and fan inside the unit.

Direct-Contact Condensers

The *direct-contact condenser* is one in which the coolant is brought into contact with the vapor. It has the advantage of low cost and simplicity of mechanical design, but its use is restricted to those applications in which mixing of the vapor and coolant is permissible.

The various types of direct contact condensers are:

1. The Spray Condenser. The coolant is sprayed, using nozzles, into a vessel to which the vapor is supplied. This is shown schematically in figure below. It is important that the spray nozzles and vessel are designed to produce a fine spray of liquid (to give a large interfacial area for heat transfer), and a long enough residence time of liquid droplets in the vessel.

2. The Baffled Column. This is similar to the spray condenser, except that the coolant is directed to flow over a series of trays in a column. The vapor is supplied to the bottom of the column. It has the advantage of countercurrent flow of vapor and coolant, though care must be taken to avoid flooding. (Flooding is an unstable condition when the vapor flow is such that the downward flow of condensate is interrupted and held up.)

3. The Packed Column. A packed column may consist of tightly-packed metal rings to increase the interfacial area for heat transfer. Liquid is supplied to the top of the column and vapor is supplied to the bottom. The disadvantage of this type of condenser is that the pressure drop is higher than in other types of direct-contact condenser.

4. The Jet Condenser. This is a device in which a jet of liquid is directed into a vapor stream, usually with the objective of desuperheating the vapor. A jet of liquid is injected into a pipeline carrying vapor via a small bore pipe and a nozzle located at the center line. The liquid is usually injected in counterflow to the vapor.

5. The Sparge Pipe. The sparge pipe consists of a pipe with holes for injecting bubbles of vapor into a pool of liquid. This is a simple method of condensing a vapor, but there are practical problems associated with generating a good distribution of bubbles of small size, which are required for efficient heat transfer.

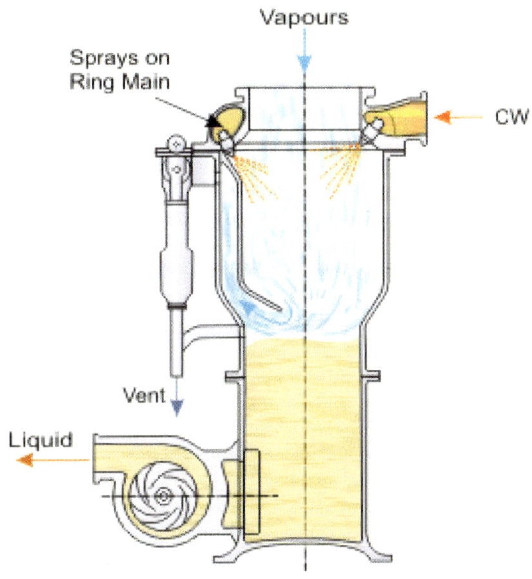

Spray condense

The design of direct-contact condensers is well described by Pair (1972). Most equipment of this type is designed on the basis of empirical information from experimental and operating data.

Copper Heat Exchanger

Copper heat exchangers are more efficient than shell and tube exchangers for low flow rates. Due to their simple construction they are low in price and easy to clean on the shell side. Their thermal efficiency approximates that of a true countercurrent flow type exchanger. Condenserors are used for condensation of vapours and cooling of liquids. Condensers are made by fusing number of parallel coils in a glass shell. Coils are made in different diameters using tubes of different bores.

Copper tubes are artificial to special requirements as to dimensional tolerances, finish and tempers for use in condensers and heat exchangers. These copper heat exchanger tubes are normally supplied in straight length in annealed & half hard temper.

Specifications

- Copper is mostly used as a metal, but when a higher hardness is required it is combined with other elements to make an alloy 5% of total use such as brass and bronze.

Applications

- A small part of copper supply is used in production of compounds for nutritional supplements and fungicides in agriculture. Machining of copper is possible, although it is usually necessary to use an alloy for intricate parts to get good machinability characteristics.
- The major applications of copper are in electrical wires (60%), roofing and plumbing (20%) and industrial machinery (15%).

Advantages

- The electrical properties of copper are exploited in copper wires and devices such as electromagnets.
- Integrated circuits and printed circuit boards increasingly feature copper in place of aluminium because of its superior electrical conductivity heat sinks and heat exchangers use copper as a result of its superior heat dissipation capacity to aluminium
- Vacuum tubes, cathode ray tubes, and the magnetrons in microwave ovens use copper, as do wave guides for microwave radiation.
- Copper's greater conductivity versus other metallic materials enhances the electrical energy efficiency of motors.

Beneficial Properties of Copper Heat Exchangers

Thermal Conductivity

Thermal conductivity (k, also denoted as λ or κ) is a measure of a material's ability to conduct heat. Heat transfer across materials of high thermal conductivity occurs at a higher rate than across materials of low thermal conductivity. In the International System of Units (SI), thermal conductivity is measured in watts per meter Kelvin (W/(m•K)). In the Imperial System of Measurement (British Imperial, or Imperial units), thermal conductivity is measured in Btu/(hr•ft·F).

Copper has a thermal conductivity of 231 Btu/(hr-ft-F). This is higher than all other metals except silver, a precious metal. Copper has a 60% better thermal conductivity rating than aluminium and a 3,000% better rating than stainless steel.

Thermal conductivity of some common metals		
Metal	Thermal conductivity	
	(Btu/(hr-ft-F))	(W/(m•K))
Silver	247.87	429
Copper	231	399
Gold	183	316
Aluminium	136	235
Yellow brass	69.33	120
Cast iron	46.33	80.1
Stainless steel	8.1	14.0

Corrosion Resistance

Corrosion resistance is essential in heat transfer applications where fluids are involved, such as in hot water tanks, radiators, etc. The only affordable material that has similar corrosion resistance to copper is stainless steel. However, the thermal conductivity of stainless steel is 1/30th that of copper. Aluminium tubes are not suitable for potable or untreated water applications because it corrodes at pH<7.0 and releases hydrogen gas.

Protective films can be applied to the inner surface of copper alloy tubes to increase corrosion resistance. For certain applications, the film is composed of iron. In power plant condensers, duplex tubes consisting of an inner titanium layer with outer copper-nickel alloys are employed. This enables the use of copper's beneficial mechanical and chemical properties (e.g., stress corrosion cracking, ammonia attack) along with titanium's excellent corrosion resistance. A duplex tube with inner aluminium brass or copper-nickel and outer stainless or mild steel can be used for cooling in the oil refining and petrochemical industries.

Biofouling Resistance

Copper and copper-nickel alloys have a high natural resistance to biofouling relative to alternative materials. Other metals used in heat exchangers, such as steel, titanium and aluminium, foul readily. Protection against biofouling, particularly in marine structures, can be accomplished over long periods of time with copper metals.

Copper-nickel alloys have been proven over many years in sea water pipework and other marine applications. These alloys resist biofouling in open seas where they do not allow microbial slime to build up and support macrofouling.

Researchers attribute copper's resistance to biofouling, even in temperate waters, to two possible mechanisms: 1) a retarding sequence of colonization through slow release of copper ions during the corrosion process, thereby inhibiting the attachment of microbial layers to marine surfaces; and/or, 2) separating layers that contain corrosive products and the larvae of macro-encrusting organisms. The latter mechanism deters the settlement of pelagic larval stages on the metal surface, rather than killing the organisms.

Antimicrobial Properties

Due to copper's strong antimicrobial properties, copper fins can inhibit bacterial, fungal and viral growths that commonly build up in air conditioning systems. Hence, the surfaces of copper-based heat exchangers are cleaner for longer periods of time than heat exchangers made from other metals. This benefit offers a greatly expanded heat exchanger service life and contributes to improved air quality. Heat exchangers fabricated separately from antimicrobial copper and aluminium in a full-scale HVAC system have been evaluated for their ability to limit microbial growth under conditions of normal flow rates using single-pass outside air. Commonly used aluminium components developed stable biofilms of bacteria and fungi within four weeks of operation. During the same time period, antimicrobial copper was able to limit bacterial loads associated with the copper heat exchanger fins by 99.99% and fungal loads by 99.74%.

Copper fin air conditioners have been deployed on buses in Shanghai to rapidly and completely kill bacteria, viruses and fungi that were previously thriving on non-copper fins and permitted to circulate around the systems. The decision to replace aluminium with copper followed antimicrobial tests by the Shanghai Municipal Center for Disease Control and Prevention (SCDC) from 2010 to 2012. The study found that microbial levels on copper fin surfaces were significantly lower than on aluminium, thereby helping to protect the health of bus passengers.

Ease of Inner Grooving

Internally grooved copper tube of smaller diameters is more thermally efficient, materially efficient, and easier to bend and flare and otherwise work with. It is generally easier to make inner grooved tubes out of copper, a very soft metal.

Plate and Frame Heat Exchanger

Plate and frame heat exchangers are used for liquid-liquid exchange at low to medium pressures. However, gasket-free plate and frame heat exchangers can safely operate at high temperatures and pressures. Plate and frame heat exchangers offer flexibility because plates can be either added or compressed for each different situation.

Plate and frame heat exchangers are made of corrugated plates on a frame. This design creates high turbulence and high wall shear stress, both of which lead to a high heat transfer coefficient and a high fouling resistance. Fluids travel within the heat exchanger. The two streams flow counter currently. The hot fluid flows down one plate while the cold fluid flows up the other plate.

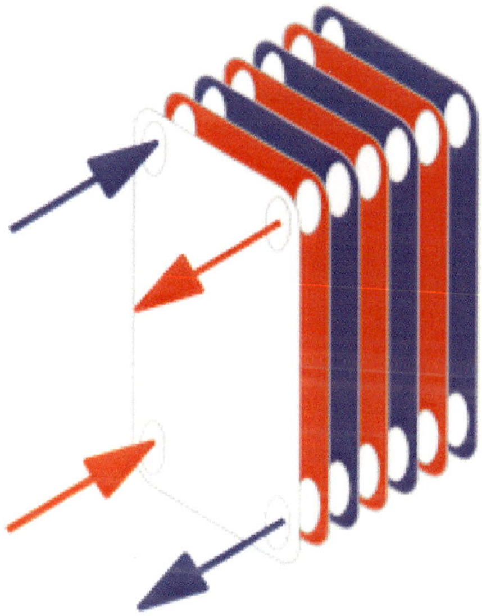

Gaskets ensure that the cold fluid and the hot fluid don't mix. Alternatives to the traditional gasket seal include brazing and laser welding.

The plates are stacked in an alternating manner to cause the counter current flow. Multiple plates are clamped together and sealed at the edges. The design allows for the two fluids to flow in alternate directions and not be mixed. However, heat can be transferred from one medium to the other through the plates.

Because gasketed plate and frame exchangers are easy to clean, they are especially useful for food and pharmaceutical processing, where high degrees of sanitation are required.

Types of Plate Heat Exchangers

There are four main types of plate heat exchangers:

- Gasketed plate heat exchangers use high quality gaskets and design to seal plates together and protect against leaks. Plates can easily be removed for cleaning, expansion, or replacing purposes, drastically reducing maintenance costs.

- Brazed Plate heat exchangers are used in many industrial and refrigeration applications. Due to stainless steel plate composition with copper brazing, they are highly resistant to corrosion. Brazed Plate heat exchangers are efficient and compact, making them an excellent economic choice.
- Welded plate heat exchangers are similar to Gasketed plate heat exchangers, but instead the plates are welded together. They are extremely durable, and are ideal for transferring fluids with high temperatures or corrosive materials. Since the plates are welded together, mechanical cleaning of plates is not an option as with plate and frame heat exchangers.
- Semi-Welded plate heat exchangers feature a mixture of welded and gasketed plates. They contain pairs of two plates welded together which are then gasketed to other pairs, therefore one fluid path is welded and the other fluid path is gasketed. This results in a plate heat exchanger that is easy to service on one side and able to transfer more intense fluids on the other. Semi-Welded heat exchangers are perfect for transferring expensive materials since they have a very low risk of fluid loss.

API Schmidt-Bretten offers all four types. Each type is suited for a number of applications in various industrial fields.

Alternatives to Plate and Frame Heat Exchangers

Plate heat exchangers are not the best choice for all applications. In situations where there is an extreme temperature difference between two fluids, it is generally more cost efficient to use a Shell & Tube heat exchanger . In a Plate heat exchanger, there can be a high pressure loss due to the large amount of turbulence created by the narrow flow channels. Applications which require a low pressure loss may want to consider a Shell & Tube heat exchanger as well.

Shell and tube exchangers consist of a number of tubes held within a shell. Heat transfer takes place between one fluid flowing through the tubes whilst another fluid flows over the tubes in the shell.

Gasketed Plate Heat Exchangers are limited in high fluid temperatures, by the temperature limitations of the gasket. Despite these limitations, Plate heat exchangers are the most efficient choice for a wide variety of applications.

Plate heat exchangers are now common and very small brazed versions are used in the hot-water sections of millions of combination boilers. The high heat transfer efficiency for such a small physical size has increased the domestic hot water flow rate of combination boilers. The small plate heat exchanger has made a great impact in domestic heating and hot water. Larger commercial versions use gasketsbetween the plates, whereas smaller versions tend to be brazed.

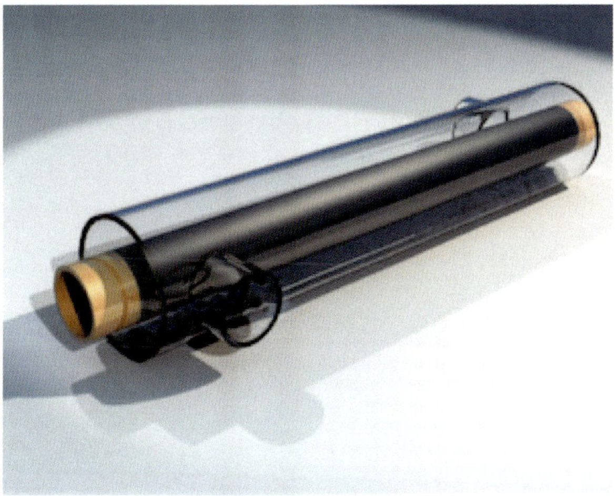

Tubular Heat Exchanger

Advantages and Disadvantages of Plate and Frame Heat Exchangers

A plate & frame heat exchanger has the following advantages over the widely used shell and tube heat exchangers:

- High value for overall heat transfer coefficient, for the same two fluids, a flat plate heat exchanger typically has a U value much higher than either a shell and tube heat exchanger or a spiral heat exchanger.
- Compact design – The combination of high value for the overall heat transfer coefficient and the general compact configuration of the flat plate heat exchanger lead to its ability to have the same thermal capacity as a shell and tube heat exchanger as much as five times its size
- Easy maintenance and cleaning – The fact that a plate and frame heat exchanger can be taken apart as discussed in the previous section, allows for easy cleaning and maintenance. A plate and frame heat exchanger can be designed to allow for easy addition or removal of plates to expand or reduce its heat transfer capacity.
- Temperature control – A flat plate heat exchanger works well with small temperature differences between the hot fluid and the cold fluid.

Plate & frame heat exchangers also has some disadvantages in comparison with other types of heat exchangers as follows:

- Potential for leakage – Although plate and frame heat exchangers are designed to allow the plates and the gaskets between them to be firmly clamped together, there is still a greater potential for leakage than with either shell and tube or spiral heat exchangers.

- Higher pressure drop – The narrow passageways for fluid flow, which lead to a high overall heat transfer coefficient, also lead to a higher pressure drop, and thus a higher cost for pumping, than shell and tube heat exchangers.
- Not good for large fluid temperature differences – A flat plate heat exchanger does not work as well as a shell and tube heat exchanger for cases where there is a large temperature difference between the two fluids.
- Doesn't work well with very high fluid temperatures – The gaskets may impose temperature limitations for plate and frame heat exchangers.

Insulating Plate and Frame Heat Exchangers

Heat exchangers must be adequately insulated to reduce heat losses. Because inspection and maintenance of heat exchangers is routine, stay-in-place insulation is not practical. Generally, heat exchanger experiences touch temperatures that differ from the ambient temperature, valuable heat is likely radiating away. In the cases of larger heat exchangers or facilities with multiple units, the amount of energy loss can be substantial.

Insulation coatings are a possible solution however it has to be reapplied each time maintenance is performed which is costly & labor intensive. Custom fit removable and reusable insulation is the most economic & efficient way to insulate heat exchangers allowing for easy on & off application as maintenance is required.

Plate & Frame Heat Exchanger Insulated

Evaluating Plate Heat Exchangers

Partially dismantled exchanger, with visible plates and gaskets

All plate heat exchangers look similar on the outside. The difference lies on the inside, in the details of the plate design and the sealing technologies used. Hence, when evaluating a plate heat exchanger, it is very important not only to explore the details of the product being supplied but also to analyze the level of research and development carried out by the manufacturer and the post-commissioning service and spare parts availability.

An important aspect to take into account when evaluating a heat exchanger are the forms of corrugation within the heat exchanger. There are two types: intermating and chevron corrugations. In general, greater heat transfer enhancement is produced from chevrons for a given increase in pressure drop and are more commonly used than intermating corrugations.

Optimization of Plate Heat Exchangers

To achieve improvement in PHE's, two important factors namely amount of heat transfer and pressure drop have to be considered such that amount of heat transfer needs to be increased and pressure drops need to be decreased. In plate heat exchangers due to presence of corrugated plate, there is a significant resistance to flow with high friction loss. Thus to design plate heat exchangers, one should consider both factors. For various range of Reynolds numbers, many correlations and chevron angles for plate heat exchangers exist. The plate geometry is one of the most important factor in heat transfer and pressure drop in plate heat exchangers, however such a feature is not accurately prescribed. In the corrugated plate heat exchangers, because of narrow path between the plates, there is a large pressure capacity and the flow becomes turbulent along the path. Therefore, it requires more pumping power than the other types of heat exchangers. Therefore, higher heat transfer and less pressure drop are targeted. The shape of plate heat exchanger is very important for industrial applications that are affected by pressure drop.

Flow Distribution and Heat Transfer Equation

Design calculations of a plate heat exchanger include flow distribution and pressure drop and heat transfer. The former is an issue of Flow distribution in manifolds. A layout configuration of plate heat exchanger can be usually simplified into a manifold system with two manifold headers for dividing and combining fluids, which can be categorized into U-type and Z-type arrangement according to flow direction in the headers, as shown in manifold arrangement. Bassiouny and Martin developed the previous theory of design. In recent years Wang unified all the main existing models and developed a most completed theory and design tool.

The total rate of heat transfer between the hot and cold fluids passing through a plate heat exchanger may be expressed as: $Q = UA\Delta Tm$ where U is the Overall heat transfer coefficient, A is the total plate area, and ΔTm is the Log mean temperature difference. U is dependent upon the heat transfer coefficients in the hot and cold streams.

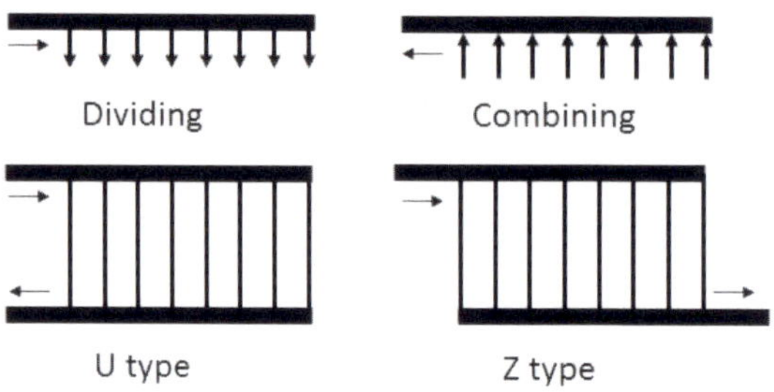

Manifold arrangement for flow distribution

Plate Fin Heat Exchanger

A plate fin heat exchanger is a form of compact heat exchanger consisting of a block of alternating layers of corrugated fins and flat separators known as parting sheets.

The concept is shown in figure. Corrugated metal fins are placed between flat plates. The structure is joined together by brazing. The fins have the dual purpose of holding the plates together, thus containing pressure, and of forming a secondary (fin) surface for heat transfer. At the edges of the plates are bars, which contain each fluid within the space between adjacent plates.

The heights of corrugations and bars may vary between plates, as shown. For a liquid stream we can use a low height corrugation, matching high heat transfer coefficient with lesser surface area while for a low pressure stream we can use a high corrugation height, matching low coefficient with higher surface area but also giving larger through area to achieve lower pressure drop. An industrial unit contains about 1000 m^2 of surface per cubic meter.

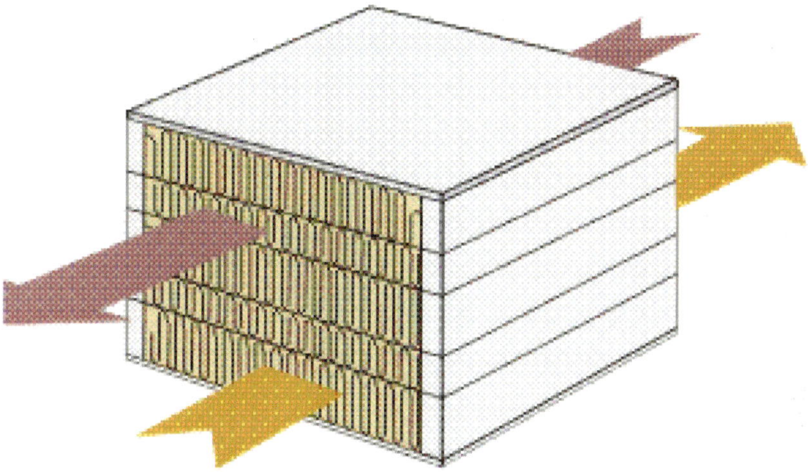

Basic construction of plate fin units

Aluminum units can be made up to 1.2 m × 1.2 m cross section and 6.2 m long.

Stainless steel units can be made up to 500 mm × 500 mm cross section × 1500 mm length.

Corrugations (Fins)

Corrugations are also made with *heat transfer enhancement devices.*

Plain corrugation is the basic form and is used normally for low pressure drop streams.

Perforated corrugation shows a slight increase in performance over plain corrugation, but this is reduced by the loss of area due to perforation. The main use is to permit migration of fluid across fin channels, usually in boiling duties.

Serrated corrugation is made by cutting the fins every 3.2 mm and displacing the second fin to a point half way between the preceding fins. This gives a dramatic increase in heat transfer.

Herringbone corrugation is made by displacing the fins sideways every 9.5 mm to give a zig-zag path. Performance is intermediate between the plain and serrated forms. The friction factor continues to fall at high Reynolds numbers, unlike the serrated, showing advantages at higher velocities and pressures.

The designer can, therefore, vary fin heights, fin pitch and fin thickness together with four standard fin types giving great versatility of design.

Construction

Plate-fin units are normally arranged for counterflow heat exchange. Cross flow units are used for vehicle radiators and cross counterflow is used for liquid subcoolers.

Figure below shows the typical layer arrangements for a *three-stream heat exchanger*. A *two-stream exchanger* can be constructed by using the first of the arrangements shown for the hot stream, alternating with the second arrangement shown for the cold stream.

In this way, the heat exchanger is built up to the appropriate height. It is then brazed together; headers and pipework are welded over the inlet and outlet parts of each stream to give a finished unit. Layers are normally arranged with alternating hot and cold streams, as below:

CHCHCHCH...

An alternative system, called double banking, is sometimes used, as below:

CHCCHCCHC...

A three-stream unit is made by using all the fin arrangements shown in figure below.

Figure below gives a drawing of a *five-stream heat exchanger*. This has "cut away" sections to show components, one of which shows the method used for stream distribution using an end entry inlet. When several cold streams are used in one unit the layers of each stream should be evenly disposed across the stack height.

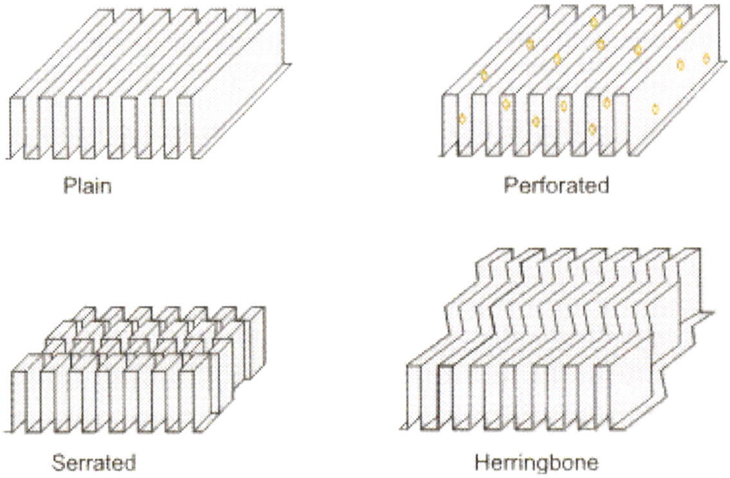

Types of corrugation

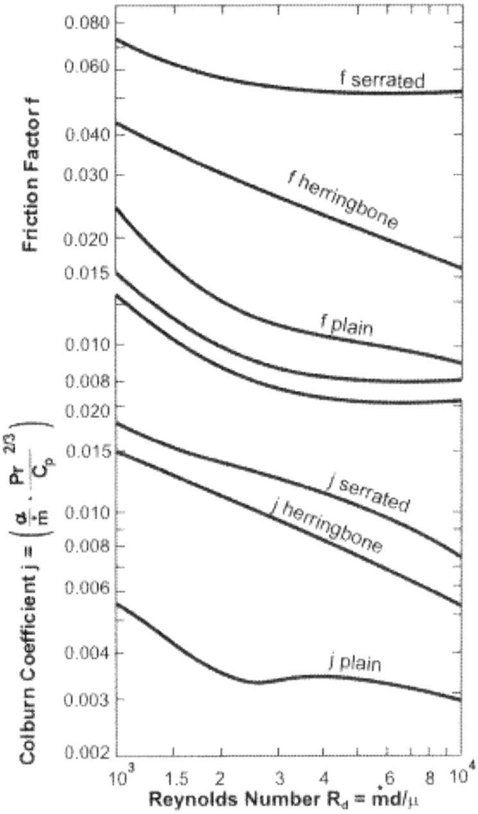

Characteristics of plate fin surfaces

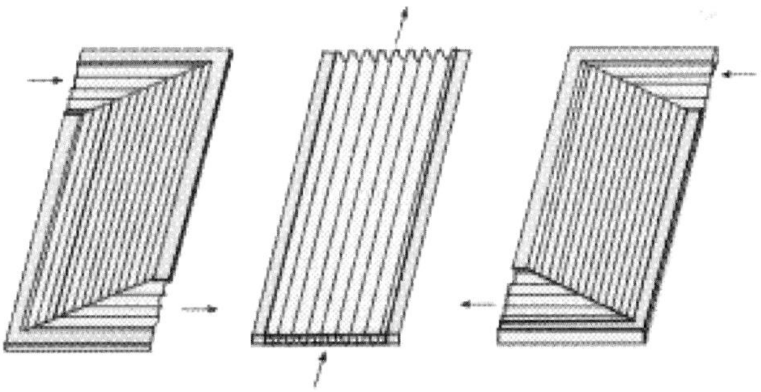

Layer arrangement for 3 stream units

Figure shows a method by which a stream which occupies layers at one end of a block can be taken out at part length and replaced by a second stream at the other end of the block. This can be repeated and in ethylene exchangers up to 5 streams occupy the same layers for different lengths.

A variation of the system allows part of a stream to be taken out (or added) at part length. The use of all these features permits a high degree of process intensification in a single block. The maximum number of streams in one block, so far, is ten, of which three had partway take-offs or additions.

Brazing and Materials

Aluminum units use material AS3003 in the exchanger block. Braze material is AS3003 + silicon. Plates are purchased with a thin film of braze metal on both sides. The unit is built and placed in a vacuum furnace. The braze takes place under vacuum and at a temperature of 580°C. The parts of the block are then firmly attached together.

AS5083 is used for headers and piping below 65°C. Above this temperature AS5454 is used.

Stainless steel units are made of AISI type 321. Braze material is essentially nickel and can be applied to the plates by spraying. Brazing takes place under vacuum at temperatures up to 1050°C.

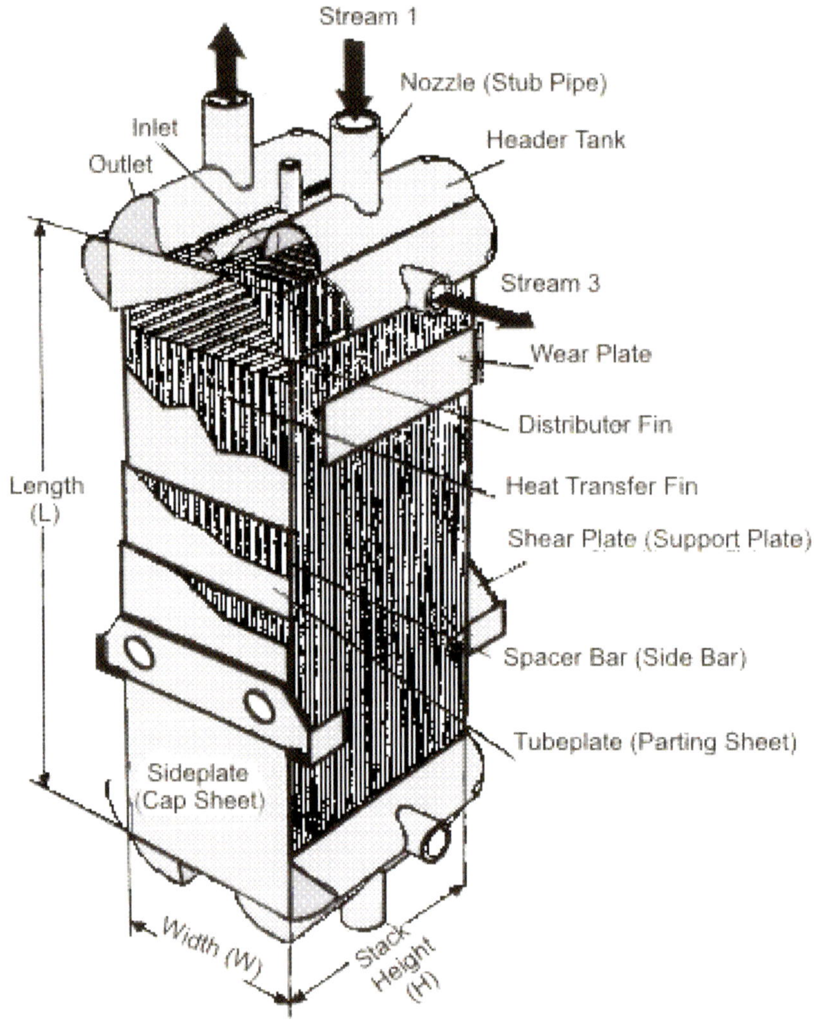

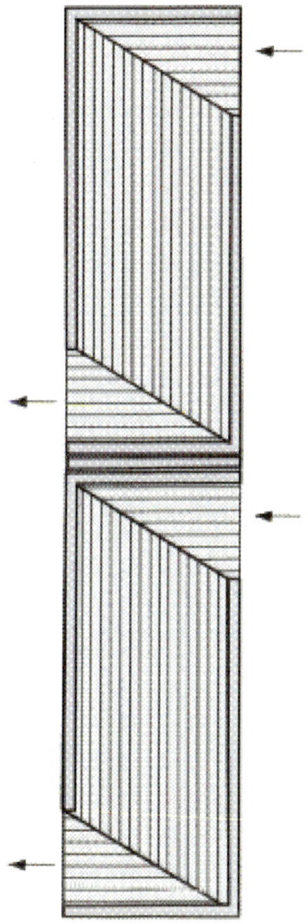

Re-entrant layers

Pressures and Temperatures

Aluminum units operate with design pressures up to 100 bars and at temperatures from absolute zcro to 65°C. Above 65°C a change of header material will allow operation to 120°C with reduced design pressures. Stainless steel units are currently limited to 50 bars design pressure and temperatures up to 750°C.

Regenerative Heat Exchanger

In recuperators, where heat is transferred directly and immediately through a partition wall of some kind, from a hot to a cold fluid, both of which flow simultaneously through the exchanger, the operation of the regenerative heat exchanger involves the temporary storage of the heat transferred in a *packing* which possesses the necessary thermal capacity. One consequence of this is that in regenerative heat exchangers or *thermal regenerators*, the hot and cold fluids pass through the same channels in the packing, alternately, both fluids washing the same surface area. In recuperators, the hot and cold fluids pass simultaneously through different but adjacent channels.

In thermal regenerator operation the hot fluid passes through the channels of the packing for a length of time called the "hot period," at the end of which, the hot fluid is switched off. A reversal now takes place when the cold fluid is admitted into the channels of the packing, initially driving out any hot fluid still resident in these channels, thereby purging the regenerator. The cold fluid then flows through the regenerator for a length of time called the "cold period," at the end of which the cold fluid is switched off and another reversal occurs in which, this time, the hot fluid purges the channels of the packing of any remaining cold fluid. A fresh hot period then begins.

During the hot period, heat is transferred from the hot fluid and is stored in the packing of the regenerator. In the subsequent cold period, this heat is regenerated and is transferred to the cold fluid passing through the exchanger.

A cycle of operation consists of a hot followed by a cold period of operation together with the necessary reversals. After many cycles of identical operation, the temperature performance of the thermal regenerator in one cycle is identical to that in the next. When this condition is realized, the heat exchanger is said to have reached "cyclic equilibrium" or "periodic steady state." Should a step change be introduced in one or more of the operating parameters, in particular, the flow rate and entrance temperature of the fluid for either period of operation, or the duration of the hot and cold periods, the regenerator undergoes a number of transient cycles until the new cyclic equilibrium is reached.

In the most common counterflow or contraflow regenerator operation, the hot gas passes through the regenerator in the opposite direction of the cold fluid. In less efficient parallel flow or co-flow the hot and cold fluids pass through the channels of the packing in the same direction. In theory, it is possible to imagine a cross-flow regenerator in which the hot and cold fluids flow in directions perpendicular to one another. This is rarely, if ever, realized in practice although cross-flow recuperators are common.

The periodic operation of regenerators can exploit the periodic operation of the system to which the exchanger is attached. For example, in hot climates, day time heat can be stored in a packing by passing the warm atmospheric air through it: this heat can then be recovered by blowing cold night time air through the same packing during the evening to provide at least some supplementary warming of the living space in a building. Hausen (1976) suggests that the throat and nasal passages act as a regenerator packing in cold weather. When an animal breathes in cold air, it is warmed as it passes through the nose and throat before the air reaches the lungs, thereby protecting the lungs from the effects of cold temperatures. As the animal breathes out, the same passages in the nose and throat are warmed by the air leaving the lungs. Clearly, the temperature of the throat and nasal tissue is also regulated by the flow of blood through it.

In general, however, a continuous supply of heated fluid is required so that the discontinuous operation of the regenerator, which is inherent in its design, must be concealed in some way.

Fixed Bed Regenerators

The most obvious technique for realizing "apparent" continuous operation, is to use two or more regenerators operating out of phase with respect to one another so that while one regenerator is supplying heated fluid, the other regenerator(s) is storing heat from the heating fluid. An apparently easy way to do this is by enclosing the set of regenerators within a system of ducts or pipes fitted with valves to facilitate the switching of the regenerators at the end of a period of operation. As one set of valves close, at a reversal, so another set open: the flow of hot gas, for example, is diverted from one regenerator to the other by the closing of such a set of valves and the opening of the other. Simultaneously, the flow of cold gas is switched from the other regenerator in a symmetric fashion. Such an arrangement is called a system *of fixed bed regenerators*, in contrast to the *rotary* regenerator which will be described shortly.

The reversal process can be more complicated than this. The following are important considerations.

- Some applications require that the regenerator(s) be purged before the supply of heated fluid, for example, is switched from one regenerator to the next. In this case, the cold period of one regenerator is extended to maintain the supply of heated fluid to the external process to which the set of regenerators is attached. Meanwhile, the hot period of the other regenerator is terminated and that regenerator is completely purged before its cold period begins. This regenerator then shoulders the burden of supplying heated fluid from the other regenerator whose end of cold period reversal can begin. Such arrangements necessarily complicate the valve and duct facilities associated with the set of regenerators: in addition, a suitable exhaust for fluids purged from the regenerator, not permitted to enter the heated fluid stream, for example, must be provided.

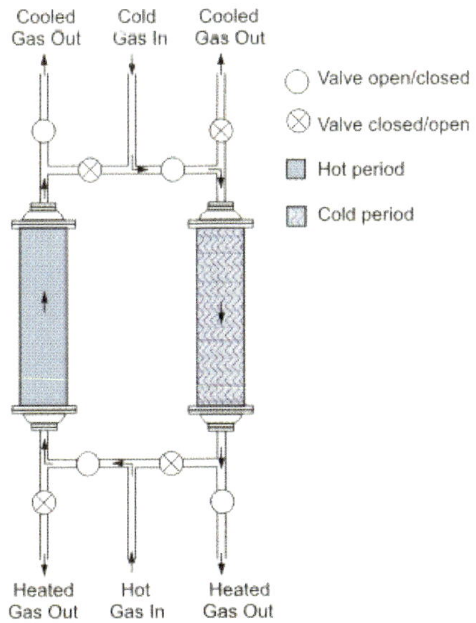

Fixed bed regenerator arrangement

- Where the fluids are gases, it is not uncommon for the pressure of the cold gas stream, for example, to be significantly higher than that of the hot gas strewn. In this case, at the end of a cold period, time must be allowed during the reversal for the regenerator to be decompressed before the hot period is permitted to begin. Similarly, time must be allowed at the start of a "cold" period for the pressure of the cold gas in the regenerator to build up before the cold period proper can begin. Again, additional valves and pipework must be provided to accommodate these complications.

In high temperature regenerators, it is desirable not to have any valves at all at the hot end of the regenerators. Where this cannot be avoided, the valves are often very expensive, perhaps requiring to be water cooled to avoid malfunction at high temperatures. It is often the case, however, the hot end of the regenerator is attached to the furnace or boiler to where pre-heated air, for the combustion of a fuel gas, is supplied directly, and from where, after a reversal, the hot gas, frequently the waste products of combustion of the fuel, is extracted directly to the regenerator. In this way, no valves are required between the regenerator and the furnace or boiler. The necessary suction of the hot gas through the regenerator is achieved by attaching the exit duct for this gas at the cold end of the regenerator to a chimney, which, if tall enough, will provide the necessary updraft. Valves are employed safely at the cold end of the regenerator to switch the heat exchanger from the chimney exhaust for waste gases to the supply of cold air for the cold period of regenerator operation, or vice-versa. A continuous supply of preheated combustion air is achieved by attaching several regenerators to a furnace or boiler, operating as necessary, out of phase with respect to one another.

Rotary Regenerators

In the *Ljungström air preheater, or rotary regenerator*, the porous packing is rotated around an axis. In its simplest form, the packing is divided into two gas tight sections and the hot and cold gases flow simultaneously in a direction parallel to this axis, usually in contra-flow, through these different segments of the packing. As the packing rotates through the hot gas stream, it stores heat, as in the hot period of a fixed bed regenerator. This thermal energy is literally transported into the cold gas stream as the packing is rotated. Once in the other gas stream, the heat is regenerated and is passed to the cold gas, as in the cold period of operation of a fixed bed system.

It is impossible to secure completely gas-tight seals at the junctions between the ducts, carrying the hot and cold fluids, respectively, and the moving surfaces of the rotating heat storing mass of the regenerator packing. However, if it is required to prevent preheated air, for example, from being contaminated by the waste combustion products which supply the necessary thermal energy, it is arranged that the pressure of the air is deliberately higher with the consequence that any leakage under the seals of the rotary regenerator is of air into the stream of hot combustion products.

When the rotor first passes from the hot gas to the cold gas stream, for example, a body of hot gas in the voids of the regenerator packing, is carried by rotation into the cold gas stream and must be purged from the regenerator, as in the fixed bed mode of operation. In some applications, it is vital that this carryover gas should not be permitted to contaminate the stream of cold gas being heated by the exchanger. In these circumstances, an additional sector is provided in the packing so that gases purged from the regenerator at the end of the hot period, for example, can be sent off to a separate exhaust, possibly fed back into the hot gas stream.

Another version of the rotary regenerator exists in which the packing remains stationary but the hoods at both ends of the packing, through which the hot and cold fluids pass in separate ducts, rotate instead.

Another type of regenerator is called a micro scale regenerative heat exchanger. It has a multilayer grating structure in which each layer is offset from the adjacent layer by half a cell which has an opening along both axes perpendicular to the flow axis. Each layer is a composite structure of two sublayers, one of a high thermal conductivity material and another of a low thermal conductivity material. When a hot fluid flows through the cell, heat from the fluid is transferred to the cell wells, and stored there. When the fluid flow reverses direction, heat is transferred from the cell walls back to the fluid.

A third type of regenerator is called a "*Rothemuhle*" regenerator. This type has a fixed matrix in a disk shape, and streams of fluid are ducted through rotating hoods. The *Rothemuhle* regenerator is used as an air preheater in some power generating plants. The thermal design of this regenerator is the same as of other types of regenerators.

Biology

We use our nose and throat as a regenerative heat exchanger when we breathe. The cooler air coming in is warmed, so that it reaches the lungs as warm air. On the way back out, this warmed air deposits much of its heat back onto the sides of the nasal passages, so that these passages are then ready to warm the next batch of air coming in. Some animals, including humans, have curled sheets of bone inside the nose called nasal turbinates to increase the surface area for heat exchange.

Cryogenics

Regenerative heat exchangers are made up of materials with high volumetric heat capacity and low thermal conductivity in the longitudinal (flow) direction. At cryogenic (very low) temperatures around 20 K, the specific heat of metals is low, and so a regenerator must be larger for a given heat load.

Advantages of Regenerators

The advantages of a regenerator over a recuperating (counter-flowing) heat exchanger is that it has a much higher surface area for a given volume, which provides a reduced exchanger volume for a given energy density, effectiveness and pressure drop. This makes a regenerator more economical in terms of materials and manufacturing, compared to an equivalent recuperator.

The design of inlet and outlet headers used to distribute hot and cold fluids in the matrix is much simpler in counter flow regenerators than recuperators. The reason behind this is that both streams flow in different sections for a rotary regenerator and one fluid enters and leaves one matrix at a time in a fixed-matrix regenerator. Furthermore, flow sectors for hot and cold fluids in rotary regenerators can be designed to optimize pressure drop in the fluids. The matrix surfaces of regenerators also have self-cleaning characteristics, reducing fluid-side fouling and corrosion. Finally properties such as small surface density and counter-flow arrangement of regenerators make it ideal for gas-gas heat exchange applications requiring effectiveness exceeding 85%. The heat transfer coefficient is much lower for gases than for liquids, thus the enormous surface area in a regenerator greatly increases heat transfer.

Disadvantages of Regenerators

The major disadvantage of rotary and fixed-matrix regenerators is that there is always some mixing of the fluid streams, and they can not be completely separated. There is an unavoidable carryover of a small fraction of one fluid stream into the other. In the rotary regenerator, the carryover fluid is trapped inside the radial seal and in the matrix, and in a fixed-matrix regenerator, the carryover fluid is the fluid that remains in the void volume of the matrix. This small fraction will mix with the other stream in the following half-cycle. Therefore, rotary and fixed-matrix regenerators are only used when it is acceptable for the two fluid streams to be mixed. Mixed flow is common for gas-to-gas heat and/or energy transfer applications, and less common in liquid or phase-changing fluids since fluid contamination is often prohibited with liquid flows.

The constant heating and cooling that takes place in regenerative heat exchangers puts a lot of stress on the components of the heat exchanger, which can cause cracking or breakdown of materials.

Mathematical Modeling of Thermal Regenerators

The relationship between the heat transferred between the fluid and the solid packing, and the heat absorbed by that packing is given by the equation

$$\bar{\alpha}A(T_f - T_s) = Mc_s \frac{\partial Ts}{\partial t}$$

On the other hand, the relationship between the thermal energy transferred between the solid packing and the fluid, and the heat taken up by the fluid passing through the regenerator is given by the equation

$$\bar{\alpha}A(T_s - T_f) = \dot{M}_f c_p L \frac{\partial T_f}{\partial t} + M_f c_p \frac{\partial T_f}{\partial t}$$

These equations apply equally to the hot and cold periods of regenerator operation, for which the relevant parameters may be different. We therefore denote the bulk heat transfer coefficient in the hot period by $\bar{\alpha}$, in the cold period by $\bar{\alpha}'$, for example.

The boundary equations relate first to the entrance gas temperature, $T_{f,in}$, where it is assumed that $T_{f,in}(t)$ = constant in each period of operation. Next, we specify that equations above consider the gas to be moving from y = 0 to y = L in both the hot and the cold periods of regenerator operation. In order to specify that the solid temperature distribution at the start of a period is equal to that at the end of the previous period, and to incorporate the contraflow operation of the regenerator, the boundary conditions are written in the form

$$T_s'(y,0) = T_s(L-y,P)$$

$$T_s(y,0) = T_s'(L-y,P')$$

where P is the duration of the hot period and P' the duration of the cold period.

The most important assumption embodied in this model is that the resistance to heat transfer at the surface of the solid and the resistance due to the finite conductivity of the packing, in a direction normal to fluid flow, can be incorporated together in a "bulk" or "lumped" heat transfer coefficient α where, using a development of [Hausen Hausen (1942)] the defining equation is given by

$$\frac{1}{\bar{\alpha}} = \frac{1}{\alpha} + \frac{\delta}{6\lambda_s}\Phi$$

where δ is the thickness of the packing and λ_δ its thermal conductivity.

The function Φ attempts to reproduce the effect of the very rapid temperature changes within the packing, immediately after a reversal, at the start of a hot or cold period. It is a function of the dimensionless parameters Ω and Ω', where

$$\Omega = \frac{K_s P}{\delta^2}, \Omega' = \frac{K_s' P'}{\delta^2}$$

In the case where the packing can be considered to be a simple, plain wall of thickness δ

$$\Phi = \frac{1}{6} - 0.00278\left\{\frac{1}{\Omega} + \frac{1}{\Omega'}\right\}$$

for

$$\frac{1}{\Omega}+\frac{1}{\Omega'}\leq 20$$

where

$$\frac{1}{\Omega}+\frac{1}{\Omega'}>20$$

then

$$\Phi=\frac{0.357}{\left(0.3+\frac{1}{2}\left(\frac{1}{\Omega}+\frac{1}{\Omega'}\right)\right)^{1/2}}$$

Similar expressions are available for the case where the packing can be regarded as a collection of solid cylinders or as a bed of spheres. Reference should be made to the later work of Hausen.

It is also assumed in this model that the thermal conductivity in a direction parallel to fluid flow, so called "longitudinal conductivity", can be neglected. This problem is discussed in the paper by Bahnke and Howard [Bahnke and Howard (1964)]. It is further idealized that the flow of fluid is uniform through the cross-section of the packing in both periods of operation. It is generally allowed, however, that the relevant thermophysical properties of both fluid and solid, including the heat transfer coefficients, can vary spatially and timewise, as a function of temperature. Equally, it is permitted to consider the case where the mass flow rates of the fluids in either or both periods of regenerator operation can vary with time.

The model is greatly simplified in the so called "linear model" in which it is further assumed that the relevant thermophysical properties of both fluid and solid, including the heat transfer coefficients, do not vary spatially and timewise, but are constant. On the other hand, they are permitted to be different between the hot and cold periods. Similarly, it is assumed that the gas flow rates are constant, although, in general $\dot{M}\neq\dot{M}'$. In these simplifying circumstances, it is possible to reduce equations above into the forms

$$\frac{\partial T_s}{\partial \eta}=T_f-T_s$$

$$\frac{\partial T_s}{\partial \xi}=T_s-T_f$$

Here, the dimensionless parameters, η for time and ξ for length, are introduced where

$$\eta=\frac{\bar{\alpha}A}{\dot{M}c_s}\left(t-\frac{M_f}{\dot{M}_fL}y\right)\ \xi=\frac{\bar{\alpha}A}{\dot{M}_fc_pL}y$$

Upon setting t = P and y = L, each period of regenerator operation is defined in terms of two dimensionless parameters given the names by Hausen [Hausen (1929)] "reduced period", Π, and "reduced length," Λ. Equations below define these for hot period operation:

$$\Pi = \frac{\bar{\alpha}A}{\dot{M}c_s}\left(P - \frac{M_f}{\dot{M}_f}\right) \quad \Lambda = \frac{\bar{\alpha}A}{\dot{M}_f c_p}$$

For the cold period, the corresponding equations are

$$\Pi' = \frac{\bar{\alpha}'A'}{M'c_s'}\left(P' - \frac{M_f'}{\dot{M}_f'}\right) \quad \Lambda' = \frac{\bar{\alpha}'A'}{\dot{M}_f' c_p'}$$

In this linear model, it is also possible to treat the hot period inlet temperature $T_{f,in} = 1.0$ and the "cold" period inlet temperature $T_{f,in} = 0$.

This is equivalent to setting

$$T_f = \frac{\tau_f - \tau'_{f,in}}{\tau_{f,in} - \tau'_{f,in}} \quad T_s = \frac{\tau_s - \tau'_{f,in}}{\tau_{f,in} - \tau'_{f,in}}$$

where τ_f and τ_s are the "real" fluid and solid temperatures and T_f and T_s are the corresponding *dimensionless* temperatures.

It is not within the orbit of this text to describe the many and varied methods of solution of these differential equations which have been developed over several years. Be it sufficient to say that there are two classes of method of solution, the "open" methods and the "closed" methods. In the open methods, a starting distribution of temperature, $T_s(\xi,0)$, of the solid at the start of a hot period is assumed. The model of the regenerator is then cycled through many cycles by solving equations above, for example, through successive hot and cold periods of operation. This simulation of the heat exchanger is then permitted to run until the periodic steady state is realized. In the closed methods, it is simply assumed that the solid temperature distribution $T_s(\xi,0)^{(n)}$, at the start of the nth cycle is equal to that, $T_s(\xi,0)^{(n+1)}$, at the start of the following cycle, at cyclic equilibrium. The equations are then solved, often in integral equation form, as a boundary-value problem. No transient cycles prior to equilibrium be established are simulated. It turns out that closed methods are useful for solving a range of linear problems but that they become extremely complicated when dealing with nonlinear problems, that is when it is permitted that the relevant thermophysical properties of both fluid and solid, including the heat transfer coefficients, can vary spatially and timewise, as a function of temperature and/or when the mass flow rates of the fluids in either or both periods of regenerator operation can vary with time. In these cases, the open methods are more easily adopted.

Construction of the Regenerator Packing

The packing of a regenerator varies considerably from one type of application to another. On the one hand, the selection of the heat storing mass is determined by the harsh, or otherwise, operating conditions under which the regenerator is required to operate. On the other, the possible operating arrangements with such packings is best understood in the context of the dimensionless parameters Λ and Π. For a given reduced length, Λ, the maximum thermal performance, it can be shown, is achieved by employing as small a value of reduced period, Π as possible.

The ratios

$$\frac{Mc_s}{\left[P - \frac{M_f}{\dot{M}_f}\right]} \text{ and } \frac{M'c_s'}{\left[P' - \frac{M_f'}{\dot{M}_f'}\right]}$$

represent the heat capacity of the packing per hot period and per cold period respectively. The effective interface between the fluid flowing through the regenerator at any instant and the heat storing packing are the products $\bar{\alpha}A$ and $\bar{\alpha}'A'$ for the hot and cold periods. The larger these interfaces, the greater must be the heat capacity per period to accommodate the thermal energy involved.

In other words, the ratios

$$\frac{\bar{\alpha}A}{Mc_s} \text{ and } \frac{\bar{\alpha}'A'}{M'c_s'}$$

must be matched by periods which yield small enough values of Π and Π' so that as good a thermal performance of the regenerator as possible can be obtained.

This can be considered in another manner: economies in regenerator size can be obtained if thin packings are used where the area to mass ratio, A/M for the hot period, A'/M' for the cold period, is large. In this case, small enough values of Π and Π' are obtained by operating the regenerator with short period lengths, that is with short cycle times.

On the other hand, harsh operating conditions might require the packing of the regenerator to be constructed of suitable materials fabricated with a robust geometrical arrangement. In this case, the area to mass ratio, A/M for the hot period, A'/M' for the cold period, might well be relatively small in which case, suitable values of Π and Π' can be obtained using long cycle times, thereby avoiding a rapid switching of the regenerators in the system used.

The matter is complicated still further if the process of reversing a regenerator is itself slow. For example, if it is necessary to pressurize the regenerator vessel at the start of the cold period, and then depressurize it at the end of the cold period, as is the case with *Cowper stoves* used for preheating the blast (of air) for iron-making, then the total cycle time must be long enough for the time necessary for these reversals not to constitute an overlarge proportion of the total cycle time. In this case, the area to mass ratio, A/M for the hot period, A'/M' for the cold period, must be forced to be small enough to generate sufficiently small values of Π and Π' with the longer hot and cold periods of operation necessary.

Very High Temperature Regenerators

Fixed bed regenerators operating with hot gas inlet temperatures in excess of 1200°K are fitted with packings constructed of fireproof refractories or ceramic materials of special quality, capable of withstanding the effects of any corrosive materials entrained in the hot gas. Thus, in glass furnace regenerators, it has not been uncommon to use high alumina packings which are capable of coping with the corrosive effects of lime, potash, silica, sodium sulfate and vanadium which can be carried over into the regenerator packing from the glass making process. In Cowper stoves, used to preheat the blast for the ironmaking and zinc smelting processes, the packing is frequently zoned: at the top of the regenerator, materials capable to withstanding the effects of very high temperatures, and further down the regenerator, of high compressive loads are used, for example, silica. At the bottom of the regenerator, it is imperative to provide materials which possess the mechanical strength and volume stability, able to support the great weight of packing above. Different silica-alumina refractories are often employed in these circumstances.

Not only must the materials of the packings be able to cope with the effects of corrosive materials, so also must the geometry of the packings be so arranged that these possibly dirty gases can have free passage through the regenerator. The blocking of the channels must be avoided. In these circumstances, various geometrical arrangements of the refractory materials must be used. "Square chimney" or "closed basket weave" arrangements are frequently used in these circumstances figure below.

It can be arranged that the channels are wide enough to provide free passage for dirty gases but for the packing not to be aligned to form chimney passages: the "open basket weave" or "staggered open basket weave" fulfill this role. The channel width can be as large as 200 mm.

Where the gases are very hot but relatively clean, as in Cowper stoves, hexagonal bricks are frequently used with passages only 50 mm wide. These passages are formed in the body and on the corners of the bricks figure below. In the *Freyn chequerwork* design, these passages are circular but other shapes are possible. These refractory bricks are arranged in layers in such a way that tubular shaped channels are formed through which the gasses can have a clear passage.

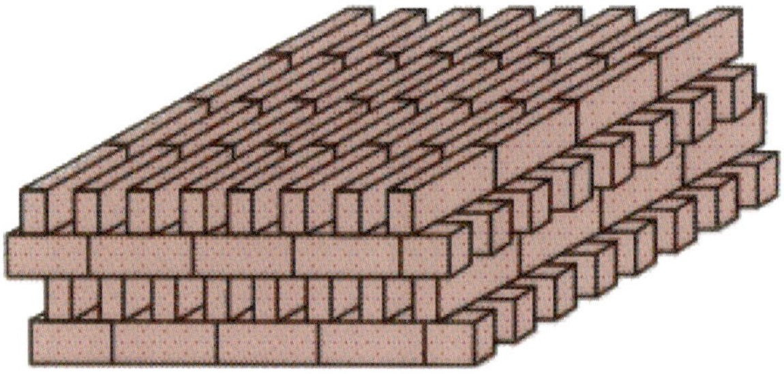

Diagram of basket weave packing arrangements

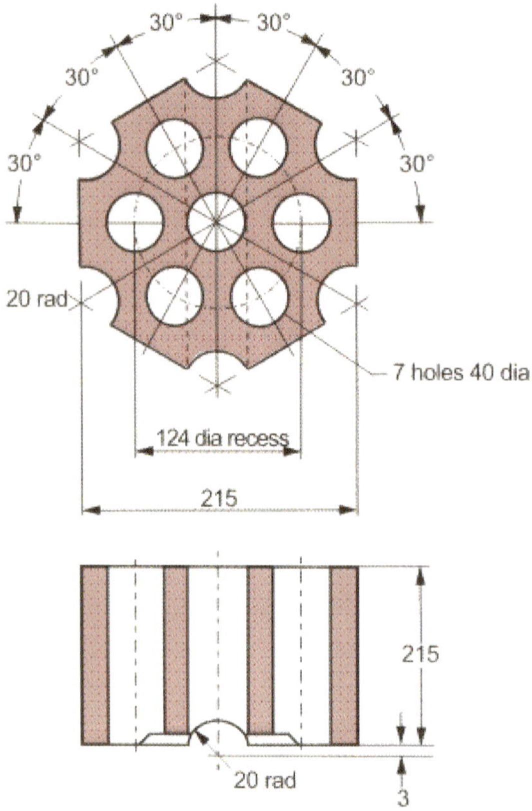

Typical design for the heat storing chequerwork for hot blast stoves (millimeter measurements)

In both kinds of arrangement, the thickness of the packing behind the available heating surface area is determined by the mechanical strength required of the packing as also by the corrosive conditions under which it is must be required to operate. Under severe conditions, thicknesses as large as 200 mm must be used; in less harsh conditions as can arise in the chemical industry, 50 mm thick bricks may be adequate.

There are two significant consequences of these possible packing arrangements. These can be understood in the context of the descriptive dimensionless parameters describing regenerators and their operations.

1. The reduced length,

$$\Lambda = \frac{\bar{\alpha}A}{\dot{M}_f}$$

is a measure of the effective heating surface area relative to the heat capacity flow rate of the hot/cold gas. Clearly, the larger the load, M_fC_p, upon the regenerator, the larger must be the heating surface area, A, to service this load. Indeed, for given operating conditions, the thermal efficiency increases with Λ. High temperature regenerators which employ basket weave or Freyn-like hexagonal packings are physically very large, in some cases, so large that they are built as part and parcel of the furnace with which they are required to operate. This is a consequence of the packing having a low surface area to volume ratio, in the range 20–30 m^{-1}. The large surface area required brings with itself a large volume of packing. Figure below gives the layout of a pair of regenerators in a Siemens arrangement, such as a glass furnace. In a Cowper stove arrangement, there are three or four regenerators held in cylindrically shaped vessels which are detached from the blast furnace they are required to serve figure below. They are still large, perhaps 30 m in height and 10 m in diameter.

2. On the other hand, the ratios

$$\frac{\bar{\alpha}A}{Mc_s} \text{ and } \frac{\bar{\alpha}'A'}{M'c_s'}$$

will be small for packings with a low surface area to volume ratio, as described above. The consequence is that small enough values of Π and Π' can be obtained using relatively long cycle times. It is not uncommon for Cowper stoves to operate with a cold period, P', of between 30 and 60 minutes and a hot period, P, of between 50 and 110 minutes for a three stove arrangement: this includes accommodating reversals lasting 5 to 10 minutes per cycle. Similarly, Siemens type furnace regenerators commonly operate with a total cycle time of about 40 minutes.

Such large regenerators are necessarily expensive. Under high temperature conditions where the gases are relatively clean, it has become common over the last ten years or so to turn to the *regenerative burner*. Here the regenerator packing consists of ceramic spheres with materials chosen to meet the operating conditions encountered. The spheres are typically 1–3 cm in diameter yielding area to volume ratios of the order of 10 times larger than encountered in the massive regenerators for the glass making furnace, for example. Area to volume ratios in the range 100–300 m^{-1} yield small compact regenerators. The particular size is determined by the load, M_fc_p, which the regenerator is required to support. A bed 0.6 m high and 0.18 m in diameter would not be uncommon, although smaller or larger beds can be used for different thermal loads.

Because the ratios

$$\frac{\bar{\alpha}A}{Mc_s} \text{ and } \frac{\bar{\alpha}'A'}{M'c_s'}$$

will be larger for beds of ceramic spheres with a surface area to volume ratio in the range 100–300 m^{-1}, it is necessary to reverse the regenerators far more rapidly than is the case for massive high temperature regenerators. The burners are operated in pairs; during the hot period hot exhaust gases are drawn from the furnace to which the burner is attached, over the bed of spheres. Simultaneously, during the cold period of the other burner fires into the furnace, using combustion air which has been preheated by the regenerative bed. The regenerators and their burners are switched after a period as short as 30 to 180 seconds. The combined burner and regenerator can be made most compact by incorporating a regenerator within the body of each burner unit. A small continuous gas fired furance, used for example in the steel industry for the annealing of strip steel, might incorporate six or more pairs of such burners. They can be operated out of phase in such a way that "apparent" continuous operation is realized by arranging that only one pair of burners is being reversed at any instant.

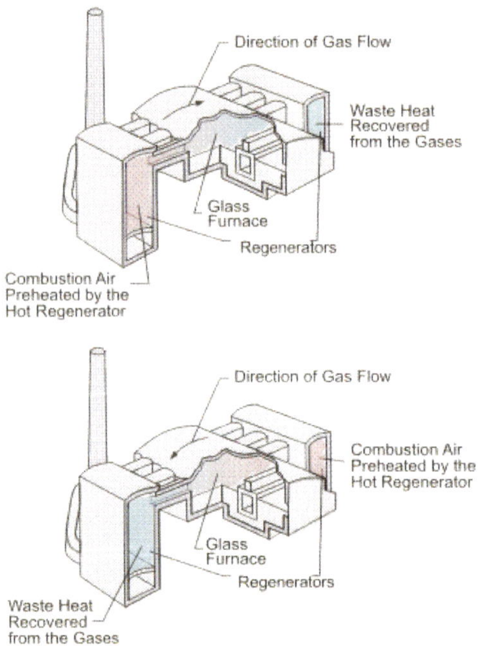

Arrangement of a pair of Siemens regenerators

Moderate Temperature Regenerators

At more moderate temperatures (400–600°C), it is common to employ rotary regenerative air preheaters. In such *Ljungström* regenerators, the cylindrical porous packing is rotated around its axis figure below. The packing materials are often fabricated of steel sheets, notched to form a large number of undulating passages. In this way, turbulent flow of the hot and cold gasses flowing through the regenerator is promoted, thereby improving the heat transfer characteristics. The metal sheets are arranged radially in removable units holding several such sheets, thereby facilitating rapid and simple maintenance.

Such metal sheeting provides a high area to volume ratio, in excess of 200 m^{-1}. Nevertheless, they must be constructed in such a way as to be able to withstand the temperatures involved as well as possibly corrosive operating conditions. Where, for example, the hot waste gases have a high SO_2 content, a vitreous enamelled heating surface can be employed to protect the steel packing at operating temperatures below the acid dew point of such gasses.

Even higher area to volume ratios can be achieved by constructing the regenerator of an assembly of sector shaped sections of a knitted mesh of wire of another material, depending on the temperature and other operating conditions. For hot gas entry temperatures of 400°C, stainless steel mesh can be employed while for temperatures of up to 800°C ceramic or alumina fibers have been considered. Other prefabricated heavy duty ceramic packings can be employed in regenerators required to withstand hot gas entry temperatures of 800°C or more: here the packing might consist of a honeycomb of ceramic material arranged as alternately flat and wave shaped layers. Such constructions realize the high surface area to volume ratio necessary to achieve compactness in regenerator construction and, at the same time, allow for a free passage for the flow of the gasses through the regenerator. They are also robust enough to survive the tough working temperatures and harsh operating conditions.

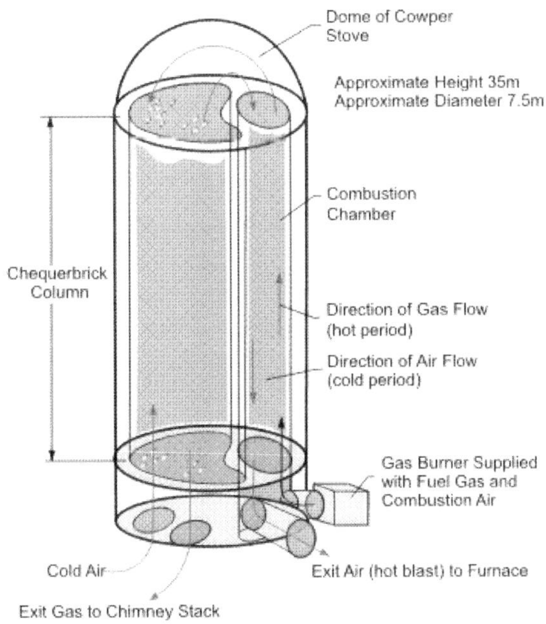

Sketch of Cowper stove used for preheating air for ironmaking blast furnace

Again the ratios

$$\frac{\bar{\alpha}A}{Mc_s} \text{ and } \frac{\bar{\alpha}'A'}{M'c_s'}$$

will be large for the rotary regenerator packings described above. It is not uncommon, therefore, for the packing to be rotated at 2–3 revolutions per minute, yielding hot/cold periods of 30 seconds or less. Small values of Π and Π' are thus generated, thereby facilitating regenerator efficiencies of 80% of more to be realized.

Lower Temperature Regenerators

The operation of regenerators at low (ambient or even lower) temperatures permits a good deal of flexibility in the choice of packing materials. Rotary regenerators for air conditioning applications employ a variety of packings which include a polyethylene terephthalate film and corrugated, knitted wire mesh. Such packings are wound round the spindle of the rotor yielding *heat wheels* of varying diameters, from 1.25 to 2.5 m. Corrugated aluminium sheets are sometimes used as are various honeycomb arrangements figure below.

A variety of packings have been developed to recover both latent heat and specific heat from one of the gases. Included in these are packings which are nonmetallic and fibrous: they can absorb moisture on the one hand but are inert to bacterial contamination on the other.

For very low temperature operation, fixed bed regenerators are frequently used where packings consist of beds of basalt or flint chips, or simply gravel. Corrugated aluminium sheets are sometimes used where corrugations run in alternate directions between the sheets which are laid on top of one another, generating fine, intersecting channels for the free passage of the gases. Such arrangements of aluminum sheets can prove to be prohibitively expensive, however.

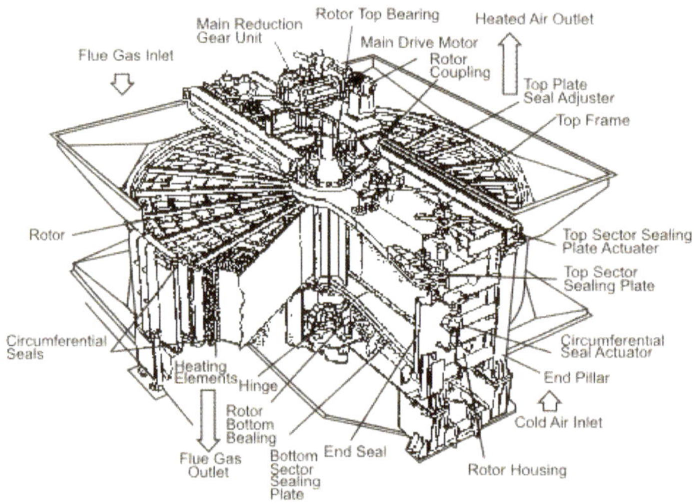

Diagram of a rotary regenerator

Honeycomb

Nomenclature

A heating surface area exposed to heating/cooling fluid m^2

a_s thermal diffusants of packing m^2/s

p specific heat, at constant pressure, of fluid J/kgK

L length of regenerator from fluid entrance to exit m

M mass of packing "behind" the surface area, A kg

M_f mass of gas resident in the voids of the packing kg

$\dot{M}_f$ mass flow rate of fluid through the regenerator kg/s

P duration of hot period s

P' duration of cold period s

T_f temperature of the fluid K

T_s temperature of the solid packing K

t time s

y distance from the regenerator entrance m

Greek Symbols

α surface heat transfer coefficient W/m^2K

$\bar{\alpha}$ bulk heat transfer coefficient W/m^2K

δ thickness of the regenerator packing m

κ_s thermal diffusivity of the packing m^2/S

λ_s thermal conductivity of the packing W/m K

Ω dimensionless duration of hot period

Ω' dimensionless duration of cold period

Dynamic Scraped Surface Heat Exchanger

Scraped surface heat exchangers (SSHEs) are commonly used in the food, chemical, and pharmaceutical industries for heat transfer, crystallization, and other continuous processes. One of the factor posing difficulties to heat transfer is viscosity. Highly viscous fluids tend to generate deep laminar flow, a condition with very poor heat transfer rates and high pressure losses involving a considerable pumping power, often exceeding the exchanger design limits. This problem becomes worsened frequently when processing non-newtonian fluids. The dynamic scraped surface heat exchangers (DSSHE) have been designed to face the above-mentioned problems. They increase heat transfer by: removing the fouling layers, increasing turbulence in case of high viscosity flow, and avoiding the generation of ice and other process by-products. The DSSHE incorporate an internal mechanism which periodically removes the product from the heat transfer wall. The product side is scraped by blades attached to a moving shaft or frame. The blades are made of a rigid plastic material to prevent damage to the scraped surface.

Depending on the arrangement of blades, there are three basic types of dynamic scraped surface heat exchangers. These include:

- Rotating/Tubular DSSHE
- Reciprocating/Tubular DSSHE
- Rotating/Plate DSSHE

Each of the three above categories serves a specific function. Our team can help you identify your needs so that we can get you a stable dynamic scraped surface heat exchanger. DSSHEs are used in a wide range of applications for many different industries. Industries that use these heaters include:

- Chemical
- Petrochemical
- Food
- Healthcare
- Pharmaceutical

Types

There are basically three types of DSSHEs depending on the arrangement of the blades:

1. *Rotating, tubular DSSHEs.* The shaft is placed parallel to the tube axis, not necessarily coincident, and spins at various frequencies, from a few dozen rpm to more than 1000 rpm. The number of blades oscillates between 1 and 4 and may take advantage of centrifugal forces to scrape the inner surface of the tube.

Examples are the Waukesha Cherry-Burrell Votator II, Alfa Laval Contherm, Terlet Terlotherm and Kelstream›s scraped surface heat exchanger.

2. *Reciprocating, tubular DSSHEs.* The shaft is concentric to the tube and moves longitudinally without rotating. The frequency spans between 10 and 60 strokes per minute. The blades may vary in number and shape, from baffle-like arrangements to perforated disk configurations. An example is the HRS Heat Exchangers Unicus.

3. *Rotating, plate DSSHEs.* The blades wipe the external surface of circular plates arranged in series inside a shell. The heating/cooling fluid runs inside the plates. The frequency is about several dozen rpm. An example is the HRS Spiratube T-Sensation.

Evaluation

Computational fluid dynamics (CFD) techniques are the standard tools to analyse and evaluate heat exchangers and similar equipment. However, for quick calculation purposes, the evaluation of DSSHEs are usually carried out with the help of adhoc (semi)empirical correlationsbased on the Buckingham π theorem

$$Fa = Fa(Re, Re', n, \ldots)$$

for pressure loss and

$$Nu = Nu(Re, Re', Pr, Fa, L/D, N, \ldots)$$

for heat transfer, where *Nu* is the Nusselt number, *Re* is the standard Reynolds number based on the inner diameter of the tube, *Re'* is the specific Reynolds number based on the wiping frequency, *Pr* is the Prandtl number, *Fa* is the Fanning friction factor, *L* is the length of the tube, *D* is the inner diameter of the tube, *n* is the number of blades and the dots account for any other relevant dimensionless parameters.

Applications

The range of applications covers a number of industries, including food, chemical, petrochemical and pharmaceutical. The DSSHEs are appropriate whenever products are prone to fouling, very viscous, particulate, heat sensitive or crystallizing.

Shell and Tube Heat Exchanger

General Description

Shell and Tube Heat Exchangers are one of the most popular types of exchanger due to the flexibility the designer has to allow for a wide range of pressures and temperatures.

There are two main categories of Shell and Tube exchanger:

1. those that are used in the petrochemical industry which tend to be covered by standards from TEMA, Tubular Exchanger Manufacturers Association;
2. those that are used in the power industry such as feedwater heaters and power plant condensers.

Regardless of the type of industry the exchanger is to be used in there are a number of common features.

A shell and tube exchanger consists of a number of tubes mounted inside a cylindrical shell. Figure 1 illustrates a typical unit that may be found in a petrochemical plant. Two fluids can exchange heat, one fluid flows over the outside of the tubes while the second fluid flows through the tubes. The fluids can be single or two phase and can flow in a parallel or a cross/counter flow arrangement.

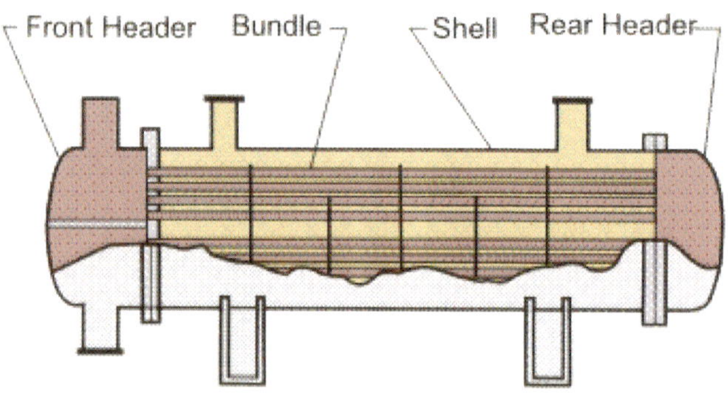

Shell and tube exchanger

The shell and tube exchanger consists of four major parts:

- Front Header—this is where the fluid enters the tubeside of the exchanger. It is sometimes referred to as the Stationary Header.
- Rear Header—this is where the tubeside fluid leaves the exchanger or where it is returned to the front header in exchangers with multiple tubeside passes.
- Tube bundle—this comprises of the tubes, *tube sheets*, *baffles* and tie rods etc. to hold the bundle together.
- Shell—this contains the *tube bundle*.

The remainder of this section concentrates on exchangers that are covered by the TEMA Standard.

Shell and Tube Exchanger: Geometric Terminology

The main components of a shell and tube exchanger are shown in figure below and described in table below.

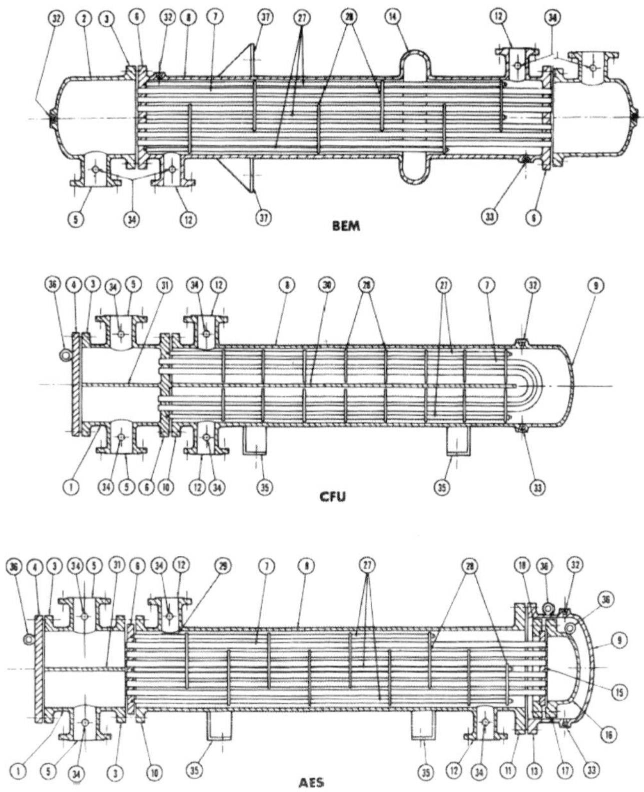

Type BEM, CFU and AES exchangers

Table: Shell and tube geometric terminology

1	Stationary (Front) Head—Channel	20	Slip-on Backing Flange
2	Stationary (Front) Head—Bonnet	21	Floating Tubesheet Skirt
3	Stationary (Front) Head Flange	22	Floating Tubesheet Skirt
4	Channel Cover	23	Packing Box Flange
5	Stationary Head Nozzle	24	Packing
6	Stationary Tubesheet	25	Packing Follower Ring
7	Tubes	26	Lantern Ring
8	Shell	27	Tie Rods and Spacers
9	Shell Cover	28	Transverse Baffles or Support Plates
10	Shell Flange—Stationary Head End	29	Impingement Baffle or Plate
11	Shell Flange—Rear Head End	30	Longitudinal Baffle
12	Shell Nozzle	31	Pass Partition
13	Shell Cover Flange	32	Vent Connection
14	Expansion Joint	33	Drain Connection
15	Floating Tubesheet	34	Instrument Connection
16	Floating Head Cover	35	Support Saddle

17	Floating Head Flange	36	Lifting Lug
18	Floating Head Backing Device	37	Support Bracket
19	Split Shear Ring		

Tema Designations

The popularity of shell and tube exchangers has resulted in a standard nomenclature being developed for their designation and use by the Tubular Exchanger Manufactures Association (TEMA). This nomenclature is defined in terms letters and diagrams. The first letter describes the front header type, the second letter the shell type and the third letter the rear header type. Figure below shows examples of a BEM, CFU, and AES exchangers while figure below illustrates the full TEMA nomenclature.

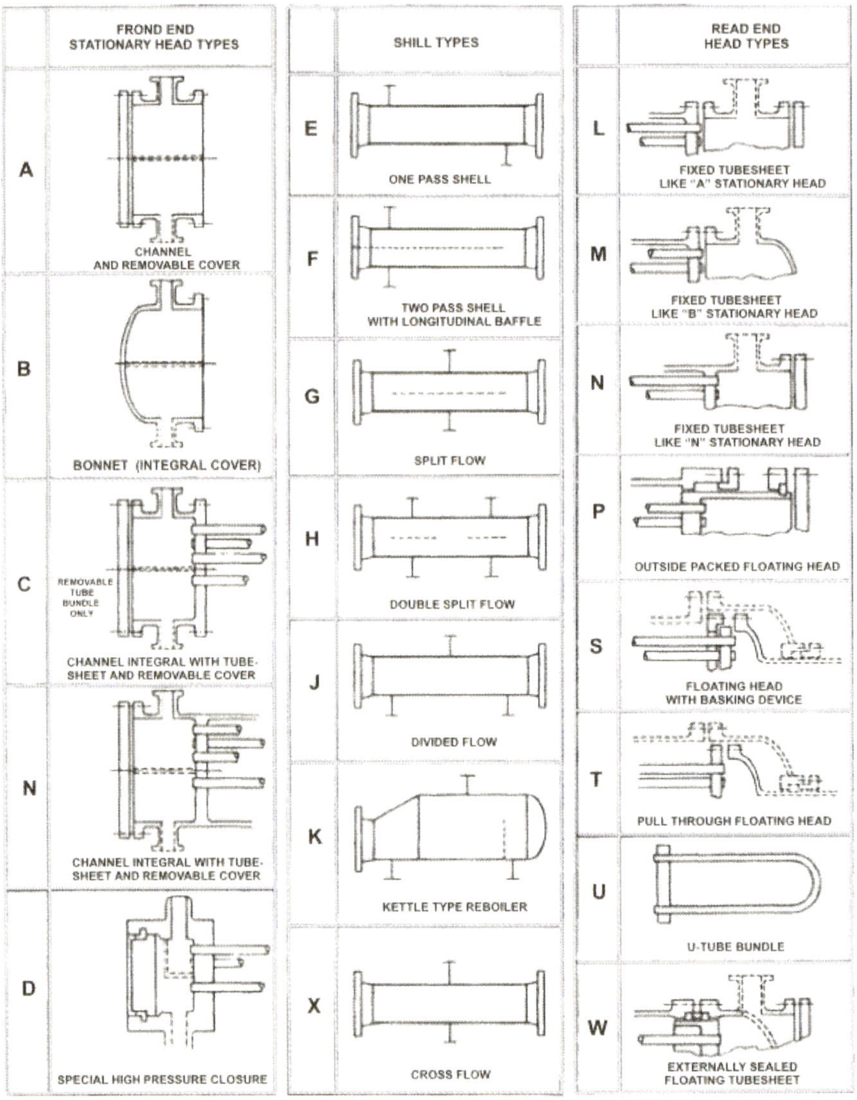

TEMA nomenclature

Many combinations of front header, shell and rear header can be made. The most common combinations for an E-Type Shell are given in table below but other combinations are also used.

Table: Shell and tube geometric terminology

Fixed tubesheet exchangers	U-tube exchangers	Floating head exchangers
AEL	AEU	AES
AEM	CEU	BES
AEN	DEU	
BEL		
BEM		
BEN		

Essentially there are three main combinations

- *Fixed tubesheet exchangers*
- *U-tube exchangers*
- *Floating header exchangers*

Fixed Tubesheet Exchanger (L, M, and N Type Rear Headers)

In a *fixed tubesheet exchanger*, the tubesheet is welded to the shell. This results in a simple and economical construction and the tube bores can be cleaned mechanically or chemically. However, the outside surfaces of the tubes are inaccessible except to chemical cleaning.

If large temperature differences exist between the shell and tube materials, it may be necessary to incorporate an expansion bellows in the shell, to eliminate excessive stresses caused by expansion. Such bellows are often a source of weakness and failure in operation. In circumstances where the consequences of failure are particularly grave U-Tube or Floating Header units are normally used.

This is the cheapest of all removable bundle designs, but is generally slightly more expensive than a fixed tubesheet design at low pressures.

U-Tube Exchangers

In a U-Tube exchanger any of the front header types may be used and the rear header is normally a M-Type. The U-tubes permit unlimited thermal expansion, the tube bundle can be removed for cleaning and small bundle to shell clearances can be achieved. However, since internal cleaning of the tubes by mechanical means is difficult, it is normal only to use this type where the tube side fluids are clean.

Floating Head Exchanger (P, S, T and W Type Rear Headers)

In this type of exchanger the tubesheet at the Rear Header end is not welded to the shell but allowed to move or float. The tubesheet at the Front Header (tube side fluid inlet end) is of a larger diameter than the shell and is sealed in a similar manner to that used in the fixed tubesheet design. The tubesheet at the rear header end of the shell is of slightly smaller diameter than the shell, allowing the bundle to be pulled through the shell. The use of a floating head means that thermal expansion can be allowed for and the tube bundle can be removed for cleaning. There are several rear header types that can be used but the S-Type Rear Head is the most popular. A floating head exchanger is suitable for the rigorous duties associated with high temperatures and pressures but is more expensive (typically of order of 25% for carbon steel construction) than the equivalent fixed tubesheet exchanger.

Considering each header and shell type in turn:

A-Type front header

This type of header is easy to repair and replace. It also gives access to the tubes for cleaning or repair without having to disturb the pipe work. It does however have two seals (one between the tube sheet and header and the other between the header and the end plate). This increases the risk of leakage and the cost of the header over a B-Type Front Header.

B-Type front header

This is the cheapest type of front header. It also is more suitable than the A-Type Front Header for high pressure duties because the header has only one seal. A disadvantage is that to gain access to the tubes requires disturbance to the pipe work in order to remove the header.

C-Type front header

This type of header is for high pressure applications (>100 bar). It does allow access to the tube without disturbing the pipe work but is difficult to repair and replace because the tube bundle is an integral part of the header.

D-Type front header

This is the most expensive type of front header. It is for very high pressures (> 150 bar). It does allow access to the tubes without disturbing the pipe work but is difficult to repair and replace because the tube bundle is an integral part of the header.

N-Type front header

The advantage of this type of header is that the tubes can be accessed without disturbing the pipe work and it is cheaper than an A-Type Front Header. However, they are difficult to maintain and replace as the header and tube sheet are an integral part of the shell.

Y-Type front header

Strictly speaking this is not a TEMA designated type but is generally recognized. It can be used as a front or rear header and is used when the exchanger is to be used in a pipe line. It is cheaper than other types of headers as it reduces piping costs. It is mainly used with single tube pass units although with suitable partitioning any odd number of passes can be allowed.

E-Type shell

This is most commonly used shell type, suitable for most duties and applications. Other shell types only tend to be used for special duties or applications.

F-Type shell

This is generally used when pure countercurrent flow is required in a two tube side pass unit. This is achieved by having two shells side passes—the two passes being separated by a longitudinal baffle. The main problem with this type of unit is thermal and hydraulic leakage across this longitudinal baffle unless special precautions are taken.

G-Type shell

This is used for horizontal thermosyphon reboilers and applications where the shell-side pressure drop needs to be kept small. This is achieved by splitting the shellside flow.

H-Type shell

This is used for similar applications to G-Type Shell but tends to be used when larger units are required.

J-Type shell

This tends to be used when the maximum allowable pressure drop is exceeded in an E-Type Shell even when double segmental baffles are used. It is also used when tube vibration is a problem. The divided flow on the shellside reduces the flow velocities over the tubes and hence reduces the pressure drop and the likelihood of tube vibration. When there are two inlet nozzles and one outlet nozzle this is sometimes referred to as an I-Type Shell.

K-Type shell

This is used only for reboilers to provide a large disengagement space in order to minimize shellside liquid carry over. Alternatively a K-Type Shell may be used as a chiller. In this case the main process is to cool the tube side fluid by boiling a fluid on the shellside.

X-Type shell

This is used if the maximum shellside pressure drop is exceeded by all other shell and baffle type combinations. The main applications are shellside condensers and gas coolers.

L-Type rear header

This type of header is for use with fixed tubesheets only, since the tubesheet is welded to the shell and access to the outside of the tubes is not possible. The main advantages of this type of header are that access can be gained to the inside of the tubes without having to remove any pipework and the bundle to shell clearances are small. The main disadvantage is that a bellows or an expansion roll are required to allow for large thermal expansions and this limits the permitted operating temperature and pressure.

M-Type rear header

This type of header is similar to the L-Type Rear Header but it is slightly cheaper. However, the header has to be removed to gain access to the inside of the tubes. Again, special measures have to be taken to cope with large thermal expansions and this limits the permitted operating temperature and pressure.

N-Type rear header

The advantage of this type of header is that the tubes can be accessed without disturbing the pipe work. However, they are difficult to maintain and replace since the header and tube sheet are an integral part of the shell.

P-Type rear header

This is an outside packed floating rear header. It is, in theory, a low cost floating head design which allows access to the inside of the tubes for cleaning and also allows the bundle to be removed for cleaning. The main problems with this type of header are:

- large bundle to shell clearances required in order to pull the bundle;
- it is limited to low pressure nonhazardous fluids, because it is possible for the shellside fluid to leak via the packing rings;
- only small thermal expansions are permitted.

In practice it is not a low cost design, because the shell has to be rolled to small tolerances for the packing to be effective.

S-Type rear header

This is a floating rear header with backing device. It is the most expensive of the floating head types but does allow the bundle to be removed and unlimited thermal expansion is possible. It also has smaller shell to bundle clearances than the other floating head types. However, it is difficult to dismantle for bundle pulling and the shell diameter and bundle to shell clearances are larger than for fixed head type exchangers.

T-Type rear header

This is a pull through floating head. It is cheaper and easier to remove the bundle than with the S-Type Rear Header, but still allows for unlimited thermal expansion. It does, however, have the largest bundle to shell clearance of all the floating head types and is more expensive than fixed header and U-tube types.

U-tube

This is the cheapest of all removable bundle designs, but is generally slightly more expensive than a fixed tubesheet design at low pressures. However, it permits unlimited thermal expansion, allows the bundle to be removed to clean the outside of the tubes, has the tightest bundle to shell clearances and is the simplest design. A disadvantage of the U-tube design is that it cannot normally have pure counterflow unless an F-Type Shell is used. Also, U-tube designs are limited to even numbers of tube passes.

W-Type rear header

This is a packed floating tubesheet with lantern ring. It is the cheapest of the floating head designs, allows for unlimited thermal expansion and allows the tube bundle to be removed for cleaning. The main problems with this type of head are:

- the large bundle to shell clearances required to pull the bundle and;
- the limitation to low pressure nonhazardous fluids (because it is possible for both the fluids to leak via the packing rings).

It is also possible for the shell and tube side fluids to become mixed if leakage occurs.

Geometric Options

Tube Diameter Layout and Pitch

Tubes may range in diameter from 12.7 mm (0.5 in) to 50.8 mm (2 in), but 19.05 mm (0.75 in) and 25.4 mm (1 in) are the most common sizes. The tubes are laid out in triangular or square patterns in the tube sheets. Figure below.

The square layouts are required where it is necessary to get at the tube surface for mechanical cleaning. The triangular arrangement allows more tubes in a given space. The tube pitch is the shortest center-to-center distance between tubes. The tube spacing is given by the tube pitch/tube diameter ratio, which is normally 1.25 or 1.33. Since a square layout is used for cleaning purposes, a minimum gap of 6.35 mm (0.25 in) is allowed between tubes.

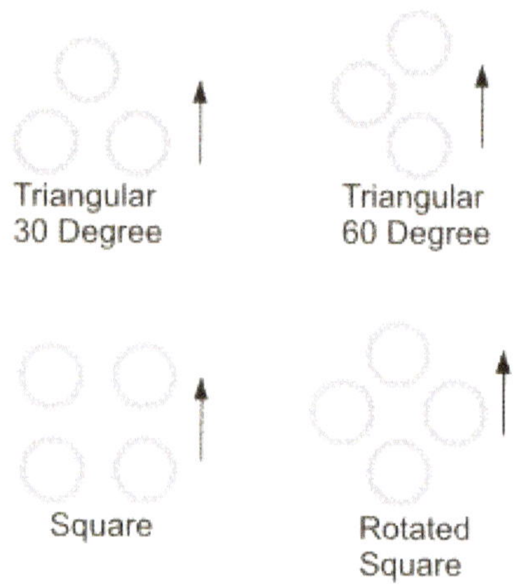

Tube layouts

Baffle Types

Baffles are installed on the shell side to give a higher heat-transfer rate due to increased turbulence and to support the tubes thus reducing the chance of damage due to vibration. There are a number of different baffle types, which support the tubes and promote flow across the tubes. Figure below shows the following baffle arrangements:

- *Single Segmental* (this is the most common),
- *Double Segmental* (this is used to obtain a lower shellside velocity and pressure drop),
- *Disc and Doughnut.*

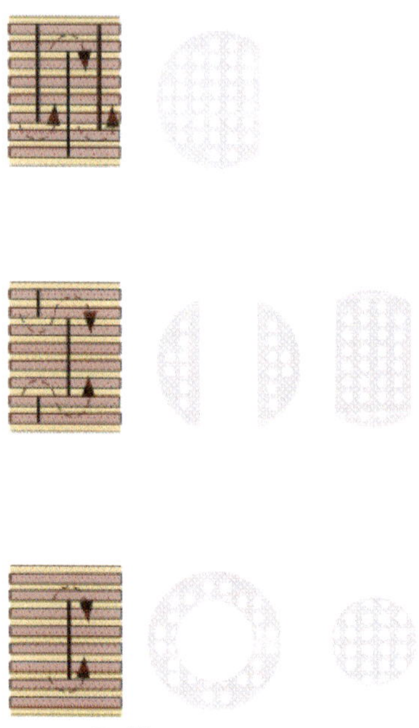

Baffle arrangements

The center-to-center distance between baffles is called the baffle-pitch and this can be adjusted to vary the crossflow velocity. In practice the baffle pitch is not normally greater than a distance equal to the inside diameter of the shell or closer than a distance equal to one-fifth the diameter or 50.8 mm (2 in) whichever is greater. In order to allow the fluid to flow backwards and forwards across the tubes part of the baffle is cut away. The height of this part is referred to as the baffle-cut and is measured as a percentage of the shell diameter, e.g., 25 per cent baffle-cut. The size of the baffle-cut (or baffle window) needs to be considered along with the baffle pitch. It is normal to size the baffle-cut and baffle pitch to approximately equalize the velocities through the window and in crossflow, respectively.

There are two main types of baffle which give longitudinal flow:

- *Orifice Baffle,*
- *Rod Baffle.*

In these types of baffle the turbulence is generated as the flow crosses the baffle.

Heat Transfer Enhancements Devices

There are three main types.

Special Surfaces

These tend to be used to promote nucleate boiling when the temperature driving force is small.

Tube Inserts

These are normally *wire wound inserts* or *twisted tapes*. They are normally used with medium to high viscosity fluids to improve heat transfer by increasing turbulence. There is also some evidence that they reduce fouling. In order to use these most effectively the exchanger should be designed for their use. This usually entails increasing the shell diameter, reducing the tube length and the number of tubeside passes in order to allow for the increased pressure loss characteristics of the devices.

Extended Surfaces

These are used to increase the heat transfer area when a stream has a low heat transfer coefficient. The most common type is *"low fin tubing"* where typically the fins are 1.5 mm high at 19 fins per inch.

Selection Criteria

In many cases the only way of ensuring optimum selection is to do a full design based on several alternative geometries. In the first instance, however, several important decisions have to be made concerning:

- allocation of fluids to the shellside and tubeside;
- selection of shell type;
- selection of front end header type;
- selection of rear end header type;
- selection of exchanger geometry.

To a large extent these often depend on each other. For instance, the allocation of a dirty fluid to the shellside directly affects the selection of exchanger tube layout.

Fluid Allocation

When deciding which side to allocate the hot and cold fluids the following need to be taken into account, in order of priority.

1. Consider any and every safety and reliability aspect and allocate fluids accordingly. Never allocate hazardous fluids such they are contained by anything other than conventional bolted and gasketted—or welded—joints.
2. Ensure that the allocation of fluids complies with established engineering practices, particularly those laid down in customer specifications.
3. Having complied with the above, allocate the fluid likely to cause the most severe mechanical cleaning problems (if any) to the tubeside.
4. If neither of the above are applicable, the allocation of the fluids should be decided only after running two alternative designs and selecting the cheapest (this is time consuming if hand calculations are used but programs such as TASC from the Heat Transfer and Fluid Flow Service (HTFS) make this a trivial task).

Shell Selection

E-type shells are the most common. If a single tube pass is used and provided there are more than three baffles, then near counter-current flow is achieved. If two or more tube passes are used, then it is not possible to obtain pure countercurrent flow and the log mean temperature difference must be corrected to allow for combined cocurrent and countercurrent flow using an F-factor.

G-type shells and H shells are normally specified only for horizontal thermosyphon reboilers. J shells and X-type shells should be selected if the allowable DP cannot be accommodated in a reasonable E-type design. For services requiring multiple shells with removable bundles, F-type shells can offer significant savings and should always be considered provided they are not prohibited by customer specifications.

Front Header Selection

The A-type front header is the standard for dirty tubeside fluids and the B-type is the standard for clean tubeside fluids. The A-type is also preferred by many operators regardless of the cleanliness of the tubeside fluid in case access to the tubes is required. Do not use other types unless the following considerations apply.

A C-type head with removable shell should be considered for hazardous tubeside fluids, heavy bundles or services requiring frequent shellside cleaning. The N-type head is used when hazardous fluids are on the tubeside. A D-type head or a B-type head welded to the tubesheet is used for high pressure applications. Y-type heads are only normally used for single tube-pass exchangers when they are installed in line with a pipeline.

Rear Header Selection

For normal service a Fixed Header (L, M, N-types) can be used provided that there is no overstressing due to differential expansion and the shellside will not require mechanical cleaning. If thermal expansion is likely a fixed header with a bellows can be used provided that the shellside fluid is not hazardous, the shellside pressure does not exceed 35 bar (500 psia) and the shellside will not require mechanical cleaning.

A U-tube unit can be used to overcome thermal expansion problems and allow the bundle to be removed for cleaning. However, countercurrent flow can only be achieved by using an F-type shell and mechanical cleaning of the tubeside can be difficult.

An S-type floating head should be used when thermal expansion needs to be allowed for and access to both sides of the exchanger is required from cleaning. Other rear head types would not normally be considered except for the special cases.

Selection of Exchanger Geometry

Tube Outside Diameter

For the process industry, 19.05 mm (3/4") tends to be the most common.

Tube Wall Thickness

Reference must be made to a recognized pressure vessel code to decide this.

Tube Length

For a given surface area, the longer the tube length the cheaper the exchanger, although a long thin exchanger may not be feasible.

TuBe Layout

45 or 90 degree layouts are chosen if mechanical cleaning is required, otherwise a 30 degree layout is often selected, because it provides a higher heat transfer and hence smaller exchanger.

Tube Pitch

The smallest allowable pitch of 1.25 times the tube outside diameter is normally used unless there is a requirement to use a larger pitch due to mechanical cleaning or tube end welding.

Number of Tube Passes

This is usually one or an even number (not normally greater than 16). Increasing the number of passes increases the heat transfer coefficient but care must be taken to ensure that the tube side ρv^2 is not greater than about 10,000 kg/m·s^2.

Shell Diameter

Standard pipe is normally used for shell diameters up to 610 mm (24"). Above this the shell is made from rolled plate. Typically shell diameters range from 152 mm to 3000 mm (6" to 120").

Baffle Type

Single segmental baffles are used by default but other types are considered if pressure drop constraints or vibration is a problem.

Baffle Spacing

This is decided after trying to balance the desire for increased crossflow velocity and tube support (smaller baffle pitch) and pressure drop constraints (larger baffle pitch). TEMA provides guidance on the maximum and minimum baffle pitch.

Baffle Cut

This depends on the baffle type but is typically 45% for single segmental baffles and 25% for double segmental baffles.

Nozzles and Impingement

For shellside nozzles the ρv^2 should not be greater than about 9000 in kg/m·s^2. For tubeside nozzles the maximum ρv^2should not exceed 2230 kg/m·s^2 for noncorrosive, nonabrasive single phase fluids and 740 kg/m·s^2 for other fluids. Impingement protection is always required for gases which are corrosive or abrasive, saturated vapors and two phases mixtures. Shell or bundle entrance or exit areas should be designed such that a ρv^2 of 5950 kg/m·s^2 is not exceeded.

Materials of Construction

In general, shell and tube exchangers are made of metal, but for specialist applications (e.g., involving strong acids or pharmaceuticals), other materials such as graphite, plastic and glass may be used.

Thermal Design

The thermal design of a shell and tube exchanger is an iterative process which is normally carried out using computer programs from organizations such as the Heat transfer and Fluid Flow Service (HTFS) or Heat Transfer Research Incorporated (HTRI). However, it is important that the engineer understands the logic behind the calculation. In order to calculate the heat transfer coefficients and pressure drops, initial decisions must be made on the sides the fluids are allocated, the front and rear header type, shell type, baffle type, tube diameter and tube layout. The tube length, shell diameter, baffle pitch and number of tube passes are also selected and these are normally the main items that are altered during each iteration in order to maximize the overall heat transfer within specified allowable pressure drops.

The main steps in the calculation are given below together with calculation methods in the open literature:

1. Calculate the shellside flow distribution.
2. Calculate the shellside heat transfer coefficient.
3. Calculate tubeside heat transfer coefficient.
4. Calculate tubeside pressure drop.
5. Calculate wall resistance and overall heat transfer coefficient.
6. Calculate mean temperature difference.
7. Calculate area required.
8. Compare area required with area of assumed geometry and allowed tubeside and shellside pressure drop with calculated values.
9. Adjust assumed geometry and repeat calculations until Area required is achieved within the allowable pressure drops.

Mechanical Design

The mechanical design of a shell and tube heat exchanger provides information on items such as shell thickness, flange thickness, etc. These are calculated using a pressure vessel design code such as the Boiler and Pressure Vessel code from ASME (American Society of Mechanical Engineers) and the British Master Pressure Vessel Standard, BS 5500. ASME is the most commonly used code for heat exchangers and is in 11 sections. Section VIII (Confined Pressure Vessels) of the code is the most applicable to heat exchangers but Sections II—Materials and Section V—Non Destructive Testing are also relevant.

Both ASME and BS5500 are widely used and accepted throughout the world but some countries insist that their own national codes are used. In order to try and simplify this the International Standards Organization is now attempting to develop a new internationally recognized code but it is likely to be a some time before this is accepted.

References

- Landes, David S. (1969). The Unbound Prometheus: Technological Change and Industrial Development in Western Europe from 1750 to the Present. Cambridge, New York: Press Syndicate of the University of Cambridge. p. 92. ISBN 0-521-09418-6
- Hindelang, Man jjhat; Palazzolo, Joseph; Robertson, Matthew, "Condensers", Encyclopedia of CHemical Engineering Equipment, University of Michigan, retrieved May 13, 2012
- Bott, T. R.; Romero, J. J. B. (October 1963). Heat Transfer Across a Scraped Surface. The Canadian Journal of Chemical Engineering. pp. 213–219
- Schmidt, Michael G.; Attaway, Hubert H.; Terzieva, Silva; Marshall, Anna; Steed, Lisa L.; Salzberg, Deborah; Hamoodi, Hameed A.; Khan, Jamil A.; Feigley, Charles E.; Michels, Harold. T. (2012). "Characterization and Control of the Microbial Community Affiliated with Copper or Aluminum Heat Exchangers of HVAC Systems". Current Microbiology. 65 (2): 141–9. doi:10.1007/s00284-012-0137-0. PMC 3378845 . PMID 22569892
- Ramesh K. Shah, Dusan P. Sekulic Fundamentals of Heat Exchanger Design,John Wiley & Sons, 2003 ISBN 0471321710
- Powell, C.A.; Preventing biofouling with copper-nickel; Copper Development Association, October 2002; "Archived copy" (PDF). Archived from the original (PDF) on 2012-10-30. Retrieved 2012-11-26

5

Mass Transfer

Mass transfer is the movement of mass from one location to another through the processes of evaporation, absorption, distillation, precipitation, etc. The topics fundamental to the study of mass transfer, such as mass transfer coefficient, mass conservation, mass balance and mass diffusivity, have been discussed in great detail in this chapter.

Mass transfer is the transport of a substance (mass) in liquid and gaseous media. Depending on the conditions, the nature, and the forces responsible for mass transfer, four basic types are distinguished: (1) diffusion in a quiescent medium, (2) mass transfer in laminar flow, (3) mass transfer in the turbulent flow, and (4) mass exchange between phases.

The simplest case is mass transfer in a medium at rest in which the driving force is the difference of concentrations in adjacent regions of the medium and the mechanism is molecular diffusion. The substance flows, by virtue of the statistical character of molecule motion, from a high concentration region to a low concentration one tending to equalization of concentration throughout the entire volume. This mass transfer is described by an equation known as Fick's law which, when applied to a binary mixture, has the form

$$\dot{m} = -D_{AB}\,\ddot{M}a\frac{dC_A}{dy},$$

where $\dot{m}$ is the flow of substance A, kg/m²s, in the reverse direction to the concentration gradient of this substance dC_A/dy (kmole/m³m), $\ddot{M}$ is the molecular weight of component A (kg/kmole) and D_{AB} (m²/s) is the interdiffusion coefficient of substance A in substance B and is determined by the physical properties of these substances. D_{AB} has been determined experimentally for many gas pairs and can also be calculated using a molecular-kinetic theory. D_{AB} is known to depend on temperature and pressure. The diffusion coefficient in gases under normal conditions is of the order of 10^{-4} m²/s, while in liquids it is about five orders lower.

Actually, in addition to concentration gradient, temperature and pressure gradients which affect mass transfer via thermal and pressure diffusion may come into play. These effects are most significant in gas mixtures with a widely varying molecule size, e.g., He—Cs. Thermal diffusion underlies one of the methods of separation of uranium isotope U^{235}.

In laminar flow of gaseous and liquid mixtures, calculation of mass transfer does not present particular difficulties. For instance, for a plate in the stream of incompressible liquid, the set of equations below describing the velocity and concentration fields has the form

$$\frac{\partial \bar{u}}{\partial x}+\frac{\partial \bar{v}}{\partial y}=0,$$

$$\bar{u}\frac{\partial \bar{u}}{\partial x}+\bar{v}\frac{\partial \bar{v}}{\partial y}=v\frac{\partial^2 \bar{u}}{\partial y^2},$$

$$\bar{u}\frac{\partial C}{\partial x}+\bar{v}\frac{\partial C}{\partial y}=D\frac{\partial^2 \bar{u}}{\partial y^2},$$

where x and y are the longitudinal and transverse coordinates, u and v the longitudinal and transverse components of velocity, v and D the coefficients of kinematic viscosity and molecular diffusion, respectively, and C the local concentration of a substance, C = f(x,y).

Under the boundary conditions $\bar{u} = \bar{v} = 0$ and $C = C_1$, at $y = 0$, $\bar{u} = \bar{u}_o$ and $C = C_0$ at $x < 0$ or $y = \infty$ the solution to equation above yields

$$Sh_x=\frac{\beta_x}{D}=0.332\,Re_x^{1/2}\,Sc^{1/3},$$

where β is the local mass transfer coefficient at the distance x from the leading edge of the plate ($\beta = [\dot{m} / \ddot{M}\,(C_1 - C_0)]$, $Re_x = x\,\bar{u}_o/v$ is the Reynolds number, and $Sc = v/D$ the Schmidt number. Note that there is a departure from the linear dependence of $\dot{m}$ or $(C_1 - C_0)$ at high rates of mass transfer.

If the total length of the plate is L, then the length-averaged mass transfer coefficient is found from the equation

$$Sh_L=\frac{\bar{\beta}L}{D}=0.664\,Re_L^{1/2}\,Sc^{1/3},$$

where $Re_L = Lu_0/v$.

Mass transfer fundamentally changes in transition to a turbulent flow. Its vortex flow characteristics lead to a large-scale transport of fluid. This transport commonly has rates which are orders of magnitude higher than molecular ones and promotes a faster equalization of the concentration field and, given a substance source, rapid propagation of the substance over the flow cross section. Since a rigorous theory of turbulence is lacking, it is desirable to describe the flow itself and heat and mass transfer in it by a set of equations similar to above equation for the laminar flow using averaged velocity values and replacing v and D by their effective values conforming to the conditions in the turbulent flow, i.e., by the coefficients of "eddy viscosity" and "turbulent diffusion". The molecular diffusion also occurs in the turbulent flow, e.g., between and inside eddies. Its role is enhanced (relative to turbulent transport) as the channel surface is approached and becomes predominant near it. It is commonly believed that the molecular D_M and turbulent D_T diffusion coefficients are additive, i.e., $D = D_M + D_T$.

Since in a developed turbulent flow the substance, energy, and momentum transport occurs via large-scale eddies, the transport rate is considered identical and D_T, a_T, and v_T are about equal. (This is a triple analogy between the transport of the substance, energy, and momentum.) This makes it possible to use empirical dimensionless equations describing heat transfer for calculation of mass transfer.

Mass exchange between a gas and a droplet is commonly encountered in engineering. For a medium at rest the solution can be written thus

$$S_c = \frac{\beta d_d}{D} = 2,$$

where β is the mass transfer coefficient, d_d the droplet diameter, and D the diffusion coefficient of the exchanged substance in the gas. When the drop moves with respect to a medium in the $Re_d \leq 200$ range the Frössling-Marshall formula

$$Sh = 2 + 0.6\,Re_d^{1/2}\,Sc^{1/3}$$

(where $Re_d = d_p u/v$ with u and v being the velocity of the gas relative to the drop and the gas kinematic viscosity, respectively) is quite consistent with the experiment. This formula also holds in the case of evaporation of droplets in a gas stream provided the evaporation rate is small or moderate.

The rate of mass transfer from the surface of liquid film (e.g. evaporation) flowing on the inner surface of the tube toward the central gas flow can be calculated using an empirical formula

$$Sh = \frac{\beta d}{D} = 0.023\,Re_d^{0.83}\,Sc^{0.44},$$

where d is the diameter of the tube (cf. Nu = 0.023 $Re_d^{0.8}\,Sc^{0.4}$ for heat transfer).

One more example of mass transfer is diffusion of some substance, such as A from one moving medium to another through interface (two-film theory). If it is assumed that there is no concentration jump on the boundary, i.e., $C_{ABi} = C_{AEi}$, then the flow of substance A can be represented in the form

$$\dot{m} = \beta_{AB} \overset{''}{M}_A (C_{AB} - C_{ABi})$$

$$= \beta_{AE} \overset{''}{M}_A (C_{AEi} - C_{AE}) = K_A \overset{''}{M}_A (C_{AB} - C_{AE}),$$

where β_{AB}, β_{AE}, and K_A are the coefficients of mass transfer for substance A in media B and E and the overall mass transfer coefficient, respectively, and C_A the concentration of substance A at the points indicated in figure below.

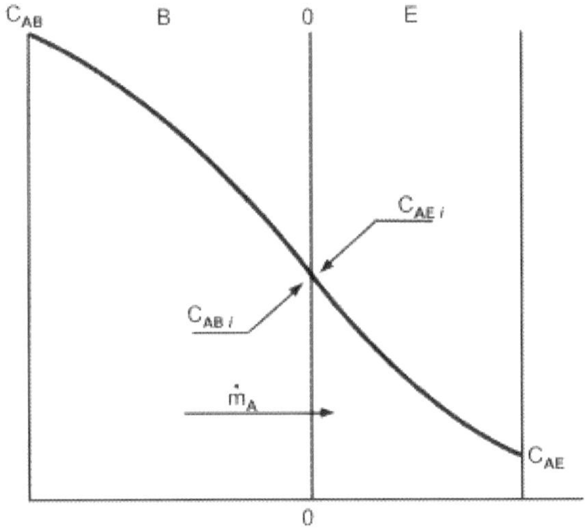

Mass transfer of component A between media B and E with no concentration jump at the interface

Hence,

$$\frac{1}{K_A} = \frac{1}{\beta_{AB}} + \frac{1}{\beta_{AC}},$$

i.e., the total resistance to mass transfer is a sum of resistances in each medium.

In many cases, the concentrations of the transferring substance are not identical at the interface in the two respective media. For instance, if B is a liquid phase and E is gas, and an equilibrium on the boundary obeys Henry's law $C_{ABi} = H\, C_{ABi}$ figure below, then

$$\dot{m} = \beta_{AB} \overset{''}{M}_A (C_{AB} - C_{ABi}) = \beta_{AE} \overset{''}{M}_A (C_{AEi} - C_{AE})$$

$$= K_A \overset{''}{M}_A (C_{AB} - HC_{AE}) = K_E \overset{''}{M}_A \left(\frac{C_{AB}}{H} - C_{AE} \right)$$

whence

$$\frac{1}{K_B} = \frac{1}{\beta_{AB}} + \frac{H}{\beta_{AC}} = \frac{H}{K_E}$$

Given the flow parameters for both films, β can be determined using, e.g., the above formulas.

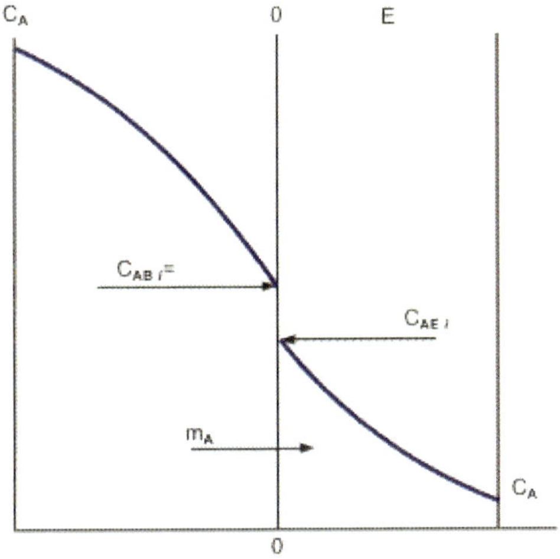

Mass transfer of component A between media B and E with a concentration jump at the interface

Mass transfer, mostly in combination with heat transfer, is widely employed in industry, in chemical process equipment, metallurgy, power engineering, and so on. The equipment includes fractionating towers, absorbers and extractors, driers and cooling towers, combustion chambers, heterogeneous catalysis apparatuses, and many others.

Mass Transfer Coefficient

Movement of the bulk fluid particles in the turbulent condition is not yet thoroughly understood. For gases it is fairly well known as a molecular diffusion since it is described in terms of kinetic theory. The rate of mass transfer from the interface to the turbulent zone in the same manner can be useful for molecular diffusion. Thus the term CD_{AB}/Z of Equation which is a characteristic of molecular diffusion is replaced by F. For binary solution,

$$N_A = \frac{N_A}{(N_A + N_B)} F \ln \left[\frac{\dfrac{N_A}{(N_A + N_B)} - y_{A_2}}{\dfrac{N_A}{(N_A + N_B)} - y_{A_1}} \right]$$

The term F is called mass transfer coefficient. The value of F depends on the local nature of the fluid motion. It is a local mass transfer coefficient defined for a particular location on the interface. Its variation depends on the effect of variation in concentrations y_{A_1} and y_{A_2} on the flux. In case of equimolar counter diffusion and transfer of one substance though stagnant another substance, special mass transfer coefficients are generally used which are defined as:

$$\text{Flux} = (\text{a coefficien t})(\text{concentration difference})$$

How mass transfer rate is calculated when there is bulk motion (turbulent) in the medium? The answer will be addressed in this study. Convective mass transfer is of two types, namely, forced convection mass transfer and free convectionmass transfer. The concept of mass transfer coefficient is to develop a simpleand practically helpful approach to convective mass transfer problems.

Mass transfer rate (W_A) is proportional to:

(i) concentration driving force (ΔC_A)

(ii) area of the contact between phases (a)

Hence, $W_A \alpha (\alpha \bullet \Delta X_A)$

$$W_A = k_C \times a \times \Delta C_A$$

where k_C is proportionality constant, called mass transfer coefficient.

Mass transfer rate, W_A can also be expressed in terms of molar flux, N_A as:

$$W_A = a \times N_A = k_C \times a \times \Delta C_A$$

$$K_C = \frac{N_A}{\Delta C_A}$$

Types of Mass Transfer Coefficients

- Mass transfer occurs in gas/liquid phase
- Choice of driving force (concentration, partial pressure, mole fraction)
- Diffusion of "A" through non-diffusing "B" / equimolar counter diffusion of "A" and "B".

Diffusion of A through non-diffusion B (N_B=0)

$$N_A = k_G (p_{A1} - p_{A2}) = k_y (y_{A1} - y_{A2}) = k_c (C_{A1} - C_{A2}) \rightarrow \text{Gas phase}$$

$$N_A = k_x (x_{A1} - x_{A2}) = k_L (C_{A1} - C_{A2}) \rightarrow \text{Liquid phase}$$

Subscripts (1 and 2) refer two positions in a medium

$$N_A = \frac{D_{AB}P(p_{A1} - P_{A2})}{RT\delta \times p_{BM}}$$

Where, $p_{BM} = \dfrac{(p_{B2} - p_{B1})}{\ln \dfrac{p_{B2}}{p_{B1}}}$ and "δ" is film thickness

Equating Equation (Gas phase) and Equation $N_A = \dfrac{D_{AB}P(p_{A1} - P_{A2})}{RT\delta \times p_{BM}}$,

$$k_G = \frac{D_{AB}P}{RT\delta \times p_{BM}}$$

Again,

$$N_A = \frac{D_{AB}P(p_{A1} - P_{A2})}{RT\delta \times p_{BM}} = \frac{D_{AB}P \times P(y_{A1} - y_{A2})}{RT\delta \times p_{BM}}$$

Equating Equation (Gas phase) and Equation $N_A = \dfrac{D_{AB}P(p_{A1} - P_{A2})}{RT\delta \times p_{BM}} = \dfrac{D_{AB}P \times P(y_{A1} - y_{A2})}{RT\delta \times p_{BM}}$,

$$k_y = \frac{D_{AB}P^2}{RT\delta \times p_{BM}}$$

Also,

$$N_A = \frac{D_{AB}P(p_{A1} - P_{A2})}{RT\delta \times p_{BM}} = \frac{D_{AB}P(C_{A1} - C_{A2})}{\delta \times p_{BM}}$$

Equating Equation (Gas phase) and Equation $N_A = \dfrac{D_{AB}P(p_{A1} - P_{A2})}{RT\delta \times p_{BM}} = \dfrac{D_{AB}P(C_{A1} - C_{A2})}{\delta \times p_{BM}}$,

$$k_C = \frac{D_{AB}P}{\delta \times p_{BM}}$$

For liquid phase diffusion,

$$N_A = \frac{D_{AB}}{\delta \times x_{BM}}\left(\frac{\rho}{M}\right)_{av}(x_{A1} - x_{A2})$$

where, $x_{BM} = \dfrac{(x_{B2} - x_{B1})}{\ln \dfrac{x_{B2}}{x_{B1}}}$ and $\left(\dfrac{\rho}{M}\right)_{av}$ is average molar concentration.

Equating Equation (Liquid phase) and Equation $N_A = \dfrac{D_{AB}}{\delta \times x_{BM}}\left(\dfrac{\rho}{M}\right)_{av}(x_{A1} - x_{A2})$,

$$k_x = \frac{D_{AB}}{\delta \times x_{BM}}\left(\frac{\rho}{M}\right)_{av}$$

Also,

$$N_A = \frac{D_{AB}}{\delta \times x_{BM}}\left(\frac{\rho}{M}\right)_{av}(x_{A1} - x_{A2}) = \frac{D_{AB}}{\delta \times x_{BM}}(C_{A1} - C_{A2})$$

Equating Equation (Liquid phase) and Equation $N_A = \frac{D_{AB}}{\delta \times x_{BM}}\left(\frac{\rho}{M}\right)_{av}(x_{A1} - x_{A2}) = \frac{D_{AB}}{\delta \times x_{BM}}(C_{A1} - C_{A2})$,

$$k_L = \frac{D_{AB}}{\delta \times x_{BM}}$$

Conversions: $k_c = RT \times k_G$; $k_y = P \times k_G$; $k_x = \left(\frac{\rho}{M}\right)_{av} k_L$.

Conservation of Mass

Conservation of mass, principle that the mass of an object or collection of objects never changes, no matter how the constituent parts rearrange themselves. Mass has been viewed in physics in two compatible ways. On the one hand, it is seen as a measure of inertia, the opposition that free bodies offer to forces: trucks are harder to move and to stop than less massive cars. On the other hand, mass is seen as giving rise to gravitational force, which accounts for the weight of an object: trucks are heavier than cars. The two views of mass are generally considered equivalent. Thus, from the perspective of either inertial mass or gravitational mass, according to the principle of mass conservation, different measurements of the mass of an object taken under various circumstances should always be the same.

With the advent of relativity theory (1905), the notion of mass underwent a radical revision. Mass lost its absoluteness. The mass of an object was seen to be equivalent to energy, to be interconvertible with energy, and to increase significantly at exceedingly high speeds near that of light. The total energy of an object was understood to comprise its rest mass as well as its increase of mass caused by high speed. The rest mass of an atomic nucleus was discovered to be measurably smaller than the sum of the rest masses of its constituent neutrons and protons. Mass was no longer considered constant, or unchangeable. In both chemical and nuclear reactions, some conversion between rest mass and energy occurs, so that the products generally have smaller or greater mass than the reactants. The difference in mass, in fact, is so slight for ordinary chemical reactions that mass conservation may be invoked as a practical principle for predicting the mass of products. Mass conservation is invalid, however, for the behaviour of masses actively involved in nuclear reactors, in particle accelerators, and in the thermonuclear reactions in the Sun and stars. The new conservation principle is the conservation of mass-energy.

Formulation and Examples

The law of conservation of mass can only be formulated in classical mechanicswhen the energy scales associated to an isolated system are much smaller than mc^2, where m is the mass of a typical object in the system, measured in the frame of reference where the object is at rest, and c is the speed of light.

The law can be formulated mathematically in the fields of fluid mechanics and continuum mechanics, where the conservation of mass is usually expressed using the continuity equation, given in differential form as

$$\frac{\partial \rho}{\partial t} + \nabla \cdot (\rho v) = 0,$$

where ρ is the density (mass per unit volume), t Is the time, $\nabla \cdot$ is the divergence, and v is the flow velocity field. The interpretation of the continuity equation for mass is the following: For a given closed surface in the system, the change in time of the mass enclosed by the surface is equal to the mass that traverses the surface, positive if matter goes in and negative if matter goes out. For the whole isolated system, this condition implies that the total mass M, sum of the masses of all components in the system, does not change in time, i.e.

$$\frac{dM}{dt} = \frac{d}{dt}\int \rho dV = 0,$$

where dV is the differential that defines the integral over the whole volume of the system.

The continuity equation for the mass is part of Euler equations of fluid dynamics. Many other convection–diffusion equations describe the conservation and flow of mass and matter in a given system.

In chemistry, the calculation of the amount of reactant and products in a chemical reaction, or stoichiometry, is founded on the principle of conservation of mass. The principle implies that during a chemical reaction the total mass of the reactant is equal to the total mass of the products. For example, in the following reaction

$$CH_4 + 2O_2 \rightarrow CO_2 + 2H_2O,$$

where one molecule of methane (CH_4) and two oxygen molecules O_2 are converted into one molecule of carbon dioxide (CO_2) and two of water (H_2O). The number of molecules as result from the reaction can be derived from the principle of conservation of mass, as initially four hydrogen atoms, 4 oxygen atoms and one carbon atom are present (as well as in the final state), then the number water molecules produced must be exactly two per molecule of carbon dioxide produced.

Many engineering problems are solved by following the mass distribution in time of a given system, this practice is known as mass balance.

Generalization

Special Relativity

In special relativity, the conservation of mass does not apply if the system is open and energy escapes. However, it does continue to apply to totally closed (isolated) systems. If energy cannot escape a system, its mass cannot decrease. In relativity theory, so long as any type of energy is retained within a system, this energy exhibits mass.

Also, mass must be differentiated from matter, since matter may *not*be perfectly conserved in isolated systems, even though mass is always conserved in such systems. However, matter is so nearly conserved in chemistry that violations of matter conservation were not measured until the nuclear age, and the assumption of matter conservation remains an important practical concept in most systems in chemistry and other studies that do not involve the high energies typical of radioactivity and nuclear reactions.

The Mass Associated With Chemical Amounts of Energy is Too Small to Measure

The change in mass of certain kinds of open systems where atoms or massive particles are not allowed to escape, but other types of energy (such as light or heat) are allowed to enter or escape, went unnoticed during the 19th century, because the change in mass associated with addition or loss of small quantities of thermal or radiant energy in chemical reactions is very small. (In theory, mass would not change at all for experiments conducted in isolated systems where heat and work were not allowed in or out.)

Mass Conservation Remains Correct if Energy is Not Lost

The conservation of relativistic mass implies the viewpoint of a single observer (or the view from a single inertial frame) since changing inertial frames may result in a change of the total energy (relativistic energy) for systems, and this quantity determines the relativistic mass.

The principle that the mass of a system of particles must be equal to the sum of their rest masses, even though true in classical physics, may be false in special relativity. The reason that rest masses cannot be simply added is that this does not take into account other forms of energy, such as kinetic and potential energy, and massless particles such as photons, all of which may (or may not) affect the total mass of systems.

For moving massive particles in a system, examining the rest masses of the various particles also amounts to introducing many different inertial observation frames (which is prohibited if total system energy and momentum are to be conserved), and also when in the rest frame of one particle, this procedure ignores the momenta of other particles, which affect the system mass if the other particles are in motion in this frame.

For the special type of mass called invariant mass, changing the inertial frame of observation for a whole closed system has no effect on the measure of invariant mass of the system, which remains both conserved and invariant (unchanging), even for different observers who view the entire system. Invariant mass is a system combination of energy and momentum, which is invariant for any observer, because in any inertial frame, the energies and momenta of the various particles always add to the same quantity (the momentum may be negative, so the addition amounts to a subtraction). The invariant mass is the relativistic mass of the system when viewed in the center of momentum frame. It is the minimum mass which a system may exhibit, as viewed from all possible inertial frames.

The conservation of both relativistic and invariant mass applies even to systems of particles created by pair production, where energy for new particles may come from kinetic energy of other particles, or from one or more photons as part of a system that includes other particles besides a photon. Again, neither the relativistic nor the invariant mass of totally closed (that is, isolated) systems changes when new particles are created. However, different inertial observers will disagree on the value of this conserved mass, if it is the relativistic mass (i.e., relativistic mass is conserved but not invariant). However, all observers agree on the value of the conserved mass if the mass being measured is the invariant mass (i.e., invariant mass is both conserved and invariant).

The mass-energy equivalence formula gives a different prediction in non-isolated systems, since if energy is allowed to escape a system, both relativistic mass and invariant mass will escape also. In this case, the mass-energy equivalence formula predicts that the *change* in mass of a system is associated with the *change* in its energy due to energy being added or subtracted: $\Delta m = \Delta E / c^2$. This form involving changes was the form in which this famous equation was originally presented by Einstein. In this sense, mass changes in any system are explained simply if the mass of the energy added or removed from the system, are taken into account.

The formula implies that bound systems have an invariant mass (rest mass for the system) less than the sum of their parts, if the binding energy has been allowed to escape the system after the system has been bound. This may happen by converting system potential energy into some other kind of active energy, such as kinetic energy or photons, which easily escape a bound system. The difference in system masses, called a mass defect, is a measure of the binding energy in bound systems – in other words, the energy needed to break the system apart. The greater the mass defect, the larger the binding energy. The binding energy (which itself has mass) must be released (as light or heat) when the parts combine to form the bound system, and this is the reason the mass of the bound system decreases when the energy leaves the system. The total invariant mass is actually conserved, when the mass of the binding energy that has escaped, is taken into account.

General Relativity

In general relativity, the total invariant mass of photons in an expanding volume of space will decrease, due to the red shift of such an expansions. The conservation of both mass and energy therefore depends on various corrections made to energy in the theory, due to the changing gravitational potential energy of such systems.

Mass Balance

The mass balance can be performed for different components of the inlet and outlet streams. When considering individual component instead of the total mass of a stream, it has to be kept in mind that for a reacting system, one or more components can be converted to one or more different components. Hence for individual component mass balance for a reacting system, we get a generation term and a consumption term. The basic mass balance for a reacting system becomes, M_N in + M_N generation = M_N out + M_N consumption + M_N accumulation M indicates mass and subscript N is second equation denotes the Nth component of the system. Hence for a system comprised N components, there are N such mass balance equations.

When equation is differentiated with respect to time, we get a mass balance equation in terms of rates. Let F denote mass flowrate, let R denote reaction rate to generate or consume a component and let A denote accumulation rate. F_N in + R_N generation = F_N out + R_N consumption + A_N

For steady state systems accumulation rate is zero and the mass balance equation is, F_N in + R_N generation = F_N out + R_N consumption.

The equations presented here are very basic mass balance equations. For specific systems each term in these equations can be detailed and solved for compositions and component mass flows in the inlet and outlet streams as well as in the system under consideration.

Illustrative Example

A simple example can illustrate the concept. Consider the situation in which a slurry is flowing into a settling tank to remove the solids in the tank. Solids are collected at the bottom by means of a conveyor beltpartially submerged in the tank, and water exits via an overflow outlet.

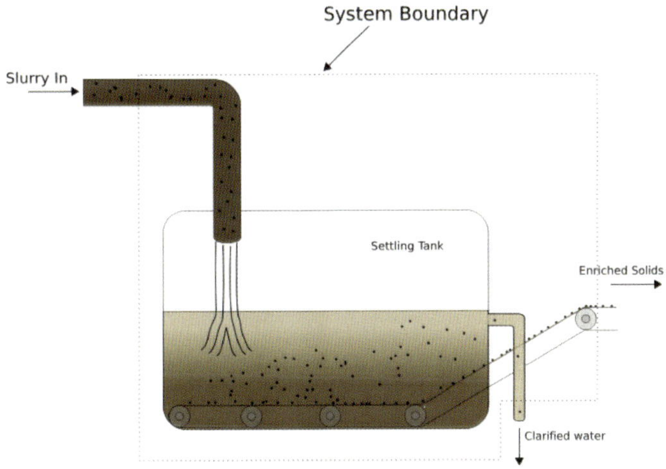

Diagram showing clarifier example

In this example, there are two substances: solids and water. The water overflow outlet carries an increased concentration of water relative to solids, as compared to the slurry inlet, and the exit of the conveyor belt carries an increased concentration of solids relative to water.

Assumptions

- Steady state
- Non-reactive system

Analysis

Suppose that the slurry inlet composition (by mass) is 50% solid and 50% water, with a mass flow of 100 kg/min. The tank is assumed to be operating at steady state, and as such accumulation is zero, so input and output must be equal for both the solids and water. If we know that the removal efficiency for the slurry tank is 60%, then the water outlet will contain 20 kg/min of solids (40% times 100 kg/min times 50% solids). If we measure the flow rate of the combined solids and water, and the water outlet is shown to be 65 kg/min, then the amount of water exiting via the conveyor belt must be 5 kg/min. This allows us to completely determine how the mass has been distributed in the system with only limited information and using the mass balance relations across the system boundaries.

Mass Feedback (Recycle)

Mass balances can be performed across systems which have cyclic flows. In these systems output streams are fed back into the input of a unit, often for further reprocessing.

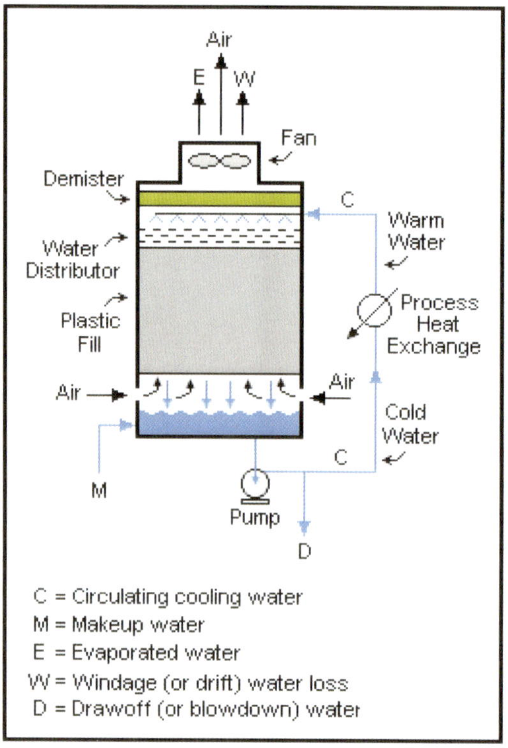

Cooling towers are a good example of a recycle system

Such systems are common in grinding circuits, where grain is crushed then sieved to only allow fine particles out of the circuit and the larger particles are returned to the roller mill (grinder). However, recycle flows are by no means restricted to solid mechanics operations; they are used in liquid and gas flows, as well. One such example is in cooling towers, where water is pumped through a tower many times, with only a small quantity of water drawn off at each pass (to prevent solids build up) until it has either evaporated or exited with the drawn off water.

The use of the recycle aids in increasing overall conversion of input products, which is useful for low per-pass conversion processes (such as the Haber process).

Differential Mass Balances

A mass balance can also be taken differentially. The concept is the same as for a large mass balance, but it is performed in the context of a limiting system (for example, one can consider the limiting case in time or, more commonly, volume). A differential mass balance is used to generate differential equations that can provide an effective tool for modelling and understanding the target system.

The differential mass balance is usually solved in two steps: first, a set of governing differential equations must be obtained, and then these equations must be solved, either analytically or, for less tractable problems, numerically.

The following systems are good examples of the applications of the differential mass balance:

1. Ideal (stirred) Batch reactor
2. Ideal tank reactor, also named Continuous Stirred Tank Reactor (CSTR)
3. Ideal Plug Flow Reactor (PFR)

Ideal Batch Reactor

The ideal completely mixed batch reactor is a closed system. Isothermal conditions are assumed, and mixing prevents concentration gradients as reactant concentrations decrease and product concentrations increase over time. Many chemistry textbooks implicitly assume that the studied system can be described as a batch reactor when they write about reaction kinetics and chemical equilibrium. The mass balance for a substance A becomes

$$IN + PROD = OUT + ACC$$

$$0 + r_A V = 0 + \frac{dn_A}{dt}$$

where r_A denotes the rate at which substance A is produced, V is the volume (which may be constant or not), n_A the number of moles (n) of substance A.

In a fed-batch reactor some reactants/ingredients are added continuously or in pulses (compare making porridge by either first blending all ingredients and then letting it boil, which can be described as a batch reactor, or by first mixing only water and salt and making that boil before the other ingredients are added, which can be described as a fed-batch reactor). Mass balances for fed-batch reactors become a bit more complicated.

Reactive Example

In the first example, we will show how to use a mass balance to derive a relationship between the percent excess air for the combustion of a hydrocarbon-base fuel oil and the percent oxygen in the combustion product gas. First, normal dry air contains 0.2095 mol of oxygen per mole of air, so there is one mole of O_2 in 4.773 mol of dry air. For stoichiometric combustion, the relationships between the mass of air and the mass of each combustible element in a fuel oil are:

$$Carbon: \frac{mass\ of\ air}{mass\ of\ C} = \frac{4.773 \times 28.96}{12.01} = 11.51$$

$$Hydrogen: \frac{mass\ of\ air}{mass\ of\ H} = \frac{\frac{1}{4}(4.773)\times 28.96}{1.008} = 34.28$$

$$Sulfur: \frac{mass\ of\ air}{mass\ of\ S} = \frac{4.773\times 28.96}{32.06} = 4.31$$

Considering the accuracy of typical analytical procedures, an equation for the mass of air per mass of fuel at stoichiometric combustion is:

$$\frac{mass\ of\ air}{mass\ of\ fuel} = \mathrm{AFR}_{mass} = 11.5(\mathrm{wC}) + 34.3(\mathrm{wH}) + (\mathrm{wS} - \mathrm{wO})$$

where wC, wH, wS, and wO refer to the mass fraction of each element in the fuel oil, sulfur burning to SO2, and AFR_{mass} refers to the air-fuel ratio in mass units.

For 1 kg of fuel oil containing 86.1% C, 13.6% H, 0.2% O, and 0.1% S the stoichiometric mass of air is 14.56 kg, so AFR = 14.56. The combustion product mass is then 15.56 kg. At exact stoichiometry, O_2 should be absent. At 15 percent excess air, the AFR = 16.75, and the mass of the combustion product gas is 17.75 kg, which contains 0.505 kg of excess oxygen. The combustion gas thus contains 2.84 percent O_2 by mass. The relationships between percent excess air and $\%O_2$ in the combustion gas are accurately expressed by quadratic equations, valid over the range 0–30 percent excess air:

$$\%\ excess\ air = 1.2804\times(\%O_2\ in\ combustion\ gas)^2 + 4.49\times(\%O_2\ in\ combustion\ gas)$$

$$\%O_2\ in\ combustion\ gas = -0.00138\times(\%\ excess\ air)^2 + 0.210\times(\%\ excess\ air)$$

In the second example we will use the law of mass action to derive the expression for a chemical equilibrium constant.

Assume we have a closed reactor in which the following liquid phase reversible reaction occurs:

$$aA + bB \leftrightarrow cC + dD$$

The mass balance for substance A becomes

$$IN + PROD = OUT + ACC$$

$$0 + r_A V = 0 + \frac{dn_A}{dt}$$

As we have a liquid phase reaction we can (usually) assume a constant volume and since $n_A = V * C_A$ we get

$$r_A V = V\frac{dC_A}{dt}$$

Or

$$r_A = \frac{dC_A}{dt}$$

In many text books this is given as the definition of reaction rate without specifying the implicit assumption that we are talking about reaction rate in a closed system with only one reaction. This is an unfortunate mistake that has confused many students over the years.

According to the law of mass action the forward reaction rate can be written as

$$r_1 = k_1[A]^a[B]^b$$

and the backward reaction rate as

$$r_{-1} = k_{-1}1[C]^c[D]^d.$$

The rate at which substance A is produced is thus.

$$r_A = a(r_{-1} - r_1)$$

and since, at equilibrium, the concentration of A is constant we get

$$r_A = a(r_{-1} - r_1) = \frac{dC_A}{dt} = 0$$

or, rearranged

$$\frac{k_1}{k_{-1}} = \frac{[C]^c[D]^d}{[A]^a[B]^b} = K_{eq}$$

Ideal Tank Reactor/Continuously Stirred Tank Reactor

The continuously mixed tank reactor is an open system with an influent stream of reactants and an effluent stream of products. A lake can be regarded as a tank reactor, and lakes with long turnover times (e.g. with low flux-to-volume ratios) can for many purposes be regarded as continuously stirred (e.g. homogeneous in all respects). The mass balance then becomes

$$IN + PROD = OUT + ACC$$

$$Q_0 \cdot C_{A,0} + r_A \cdot V = Q \cdot C_A + \frac{dn_A}{dt}$$

where Q_o and Q denote the volumetric flow in and out of the system respectively and $C_{A,o}$ and C_A the concentration of A in the inflow and outflow respective. In an open system we can never reach a chemical equilibrium. We can, however, reach a steady state where all state variables (temperature, concentrations etc.) remain constant $ACC = 0$.

Example

Consider a bathtub in which there is some bathing salt dissolved. We now fill in more water, keeping the bottom plug in. What happens?

Since there is no reaction, $PROD = 0$ and since there is no outflow $Q = 0$. The mass balance becomes

$$IN + PROD = OUT + ACC$$

$$Q_0 \cdot C_{A,0} + 0 = 0 \cdot C_A + \frac{dn_A}{dt}$$

Or

$$Q_0 \cdot C_{A,0} = \frac{dC_A V}{dt} = V\frac{dC_A}{dt} + C_A\frac{dV}{dt}$$

Using a mass balance for total volume, however, it is evident that $\frac{dV}{dt} = Q_0$ and that $V = V_{t=0} + Q_0 t$. Thus we get

$$\frac{dC_A}{dt} = \frac{Q_0}{(V_{t=0} + Q_0 t)}\left(C_{A,0} - C_A\right)$$

Note that there is no reaction and hence no reaction rate or rate law involved, and yet $\frac{dC_A}{dt} \neq 0$. We can thus draw the conclusion that reaction rate can not be defined in a general manner using $\frac{dC}{dt}$. One must first write down a mass balance before a link between $\frac{dC}{dt}$ and the reaction rate can be found. Many textbooks, however, define reaction rate as

$$v = \frac{dC_A}{dt}$$

without mentioning that this definition implicitly assumes that the system is closed, has a constant volume and that there is only one reaction.

Ideal Plug Flow Reactor (PFR)

The idealized plug flow reactor is an open system resembling a tube with no mixing in the direction of flow but perfect mixing perpendicular to the direction of flow. Often used for systems like rivers and water pipes if the flow is turbulent. When a mass balance is made for a tube, one first considers an infinitesimal part of the tube and make a mass balance over that using the ideal tank reactor model. That mass balance is then integrated over the entire reactor volume to obtain:

$$\frac{d(Q \cdot C_A)}{dV} = r_A$$

In numeric solutions, e.g. when using computers, the ideal tube is often translated to a series of tank reactors, as it can be shown that a PFR is equivalent to an infinite number of stirred tanks in series, but the latter is often easier to analyze, especially at steady state.

More Complex Problems

In reality, reactors are often non-ideal, in which combinations of the reactor models above are used to describe the system. Not only chemical reaction rates, but also mass transfer rates may be important in the mathematical description of a system, especially in heterogeneous systems.

As the chemical reaction rate depends on temperature it is often necessary to make both an energy balance (often a heat balance rather than a full-fledged energy balance) as well as mass balances to fully describe the system. A different reactor model might be needed for the energy balance: A system that is closed with respect to mass might be open with respect to energy e.g. since heat may enter the system through conduction.

Mass Diffusivity

Diffusion coefficient is the proportionality factor D in *Fick's law* by which the mass of a substance dM diffusing in time dt through the surface dF normal to the diffusion direction is proportional to the concentration gradient grad c of this substance:

dM = −D grad c dF dt. Hence, physically, the diffusion coefficient implies that the mass of the substance diffuses through a unit surface in a unit time at a concentration gradient of unity. The dimension of D in the SI system is a square meter per second.

The diffusion coefficient is a physical constant dependent on molecule size and other properties of the diffusing substance as well as on temperature and pressure. Diffusion coefficients of one substance into the other are commonly determined experimentally and presented in reference tables. Here, examples of self-diffusion and interdiffusion (binary diffusion) coefficients in a gaseous and liquid media are given in tables below.

Table: Self-diffusion coefficient D_A of some gases at T = 273 K and p = 0.1 MPa

Gas	DA, cm²/s	Gas	DA, cm2/s	Gas	DA, cm²/s
H_2	1.604	O_2	0.192	H_2O	0.276
He	1.386	N_2	0.155	CH_4	0.188
Ar	0.157	CO_2	0.106	NH_3	0.192

Table: Interdiffusion coefficient D_{AB}

Diffusing gas A	Medium B	P. mmH_2O	T,K	DAB. Cm²/s
H_2	He	760	298	1.64
Ar	He	751	288	0.703
N_2	CO_2	760	288	0.158
H_2O	air	760	273	0.219
H_2	N_2	760	288	0.743
H_2O	CO_2	760	273	0.146

Table: Diffusion coefficient of gases in liquids

Gas	Liquid	T. K	D,10−5cm/s
Air	water	293	2.5
CO_2	water	298	1.92
H_2	water	298	4.50
NH_3	water	285	1.64
CO_2	ethanol	298	3.42
CO_2	heptane	298	6.03

As is obvious from comparing the data of tables above, the diffusion coefficients in a gaseous and a liquid phases differ by a factor of $10^4 - 10^5$, which is quite reasonable considering that diffusion is the movement of individual molecules through the layer of molecules of the same substance (self-diffusion) or other substances (binary diffusion in which the molecules of two substances interdiffuse). The number density of molecules in liquid is also very much higher and their mobility is lower, which implies a much lower diffusion coefficient In solids, diffusion is still slower.

If experimental data are lacking, the diffusion coefficient can be calculated.

Diffusion in gases. For ideal gases, the diffusion coefficient does not depend on substance concentration. In accordance with the kinetic theory of gases, the mean free-path length l of molecules is inversely proportional to the mean cross-sectional area of the molecule S and the number density of the molecules n in a mixture. The latter is inversely proportional to the space occupied by the mixture, i.e., T/p, where T is the temperature and p, the pressure. The mean velocity of the molecules u is proportional to $(T/\bar{\mu})^{1/2}$, where $\bar{\mu}$ is the molecular mass. Thus, in the case of interdiffusion of gases with the same molecular mass μ, or self-diffusion, the expression is:

$$D - ul - \frac{1}{n}\sqrt{\frac{T}{\bar{\mu}}} - \frac{T^{3/2}}{pS\bar{\mu}^{1/2}}.$$

If molecules of types A and B interact (binary diffusion), then the interdiffusion coefficient is:

$$D_{AB} = \text{const}\frac{T^{3/2}\left(1/\bar{\mu}_A + 1/\bar{\mu}_B\right)^{1/2}}{pS_{AB}}$$

where S_{AB} is the mean value of the cross-sections of molecules of both types. The kinetic theory of gases makes it possible to determine the constant in equation above, assuming the molecules are spherical and their cross-sections are equal to cross-sections of these spheres.

Sutherland has made a correction to equation above, taking into account the forces of intermolecular attraction which influence the free-path length of molecules; thus,

$$D_{AB} = \text{const}\frac{T^{3/2}}{pS_{AB}}\left(\frac{1}{\bar{\mu}_A} + \frac{1}{\bar{\mu}_B}\right)^{1/2}\frac{1}{1+\frac{C_{AB}}{T}}$$

where C_{AB} is the *Sutherland coefficient.*

Contemporary kinetic theory takes into account the intricate character of molecular interaction; molecules repel one another when they are close and attract one another at a distance. Many researchers have studied the potentials of this molecular interaction, but great recognition has been won by the so-called *Lennard-Jones potential*

$$\varphi(r) = 4\varepsilon\left[\left(\frac{\sigma}{r}\right)^{12} - \left(\frac{\sigma}{r}\right)^{6}\right],$$

where φ(r) is the potential energy; r, the distance between the centers of molecules; ε and σ are Lennard-Jones' interaction constants determined for many gases and summarized in tables.

For dilute gas mixtures, and assuming that the molecular collisions are only binary and elastic, that the motion of colliding molecules is in terms of classical mechanics, that quantum effects are absent, and, finally, intermolecular forces act only along the center line, the expression

$$D_{AB} = \frac{1.883\times10^{-20}T^{3/2}\left(1/\bar{\mu}_A + 1/\bar{\mu}_B\right)^{1/2}}{p\sigma_{AB}^2\Omega}$$

for the diffusion coefficient in a binary mixture is obtained by Bird, Hirshfelder and Curtiss. Here, $\Omega = f(kT/\varepsilon_{AB})$ is the collision integral and k is the Boltzmann's constant. The interaction parameters ε_{AB} and σ_{AB} determined for the binary system from the appropriate constants for pure substances:

$$\frac{\varepsilon_{AB}}{k} = \left(\frac{\varepsilon_A}{k}\cdot\frac{\varepsilon_B}{k}\right)^{1/2}, \sigma_{AB} = \frac{\sigma_A + \sigma_B}{2}$$

If the data on ε and σ are not available, they can be estimated using the well-known critical parameters for a given substance:

$$\frac{\varepsilon}{k}0.75\,T_c, \sigma = \frac{5}{6}\cdot10^{-8}V_c^{1/3},$$

where T_c and V_c are the critical temperature (K) and the critical molar volume (cm³/mol), respectively.

Wilke and Lee noted the coefficient 1.885Ч10⁻² is not constant in reality and depends on molecular masses of the diffusing gases:

$$\left[1.533 + 0.352\left(\frac{\bar{\mu}_A + \bar{\mu}_B}{\bar{\mu}_A\cdot\bar{\mu}_B}\right)^{0.5}\cdot10^{-20}\right]$$

This accurate definition results in a better agreement between experimental and computed data.

A correlation formula obtained by *Fuller, Schetter and Gittings* by means of computer-aided correlation of 340 experimental points, expressed as:

$$D_{AB} = \frac{1.0110^{-4}T^{1.75}\left(\left(1/\bar{\mu}_A + 1/\bar{\mu}_B\right)^{1/2}\right)}{p\left[\left(\sum V_A\right)^{1/3} + \left(\sum V_B\right)^{1/3}\right]^2}$$

has come to be widely known, where ($\sum V_A$) and ($\sum V_B$) are the values derived from summation of atomic diffusion volumes for each component of the binary mixture, i.e., molecules A and B. The values for some atoms and simple molecules are presented in table above.

Table: Diffusion volumes of atoms and simple molecules

Substance	$\bar{v}$	Substance	$\bar{v}$	Substance	$\bar{v}$
C	16.5	CO2	26.9	CO	18.9
H	1.98	H_2	7.07	NH_3	14.9
O	5.48	O_2	16.6	H_2O	12.7
N	5.69	N_2	17.9	SO_2	41.1

In order to calculate the diffusion coefficient in multicomponent systems, Wilke used the Maxwell-Stefan equation to derive the expression

$$D'_A = \frac{1-Y_A}{Y_B/D_{AB} + Y_C/D_{AC} + ...},$$

where D'_A is the diffusion coefficient of the component A in the mixture with B, C,..., Y_A, Y_B, Y_C are the molar fractions of the appropriate components; and D_{AB} and D_{AC} are the diffusion coefficients in the AB and AC binary systems, respectively.

Diffusion in liquids. As has been noted, diffusion in liquids encounters greater resistance and the diffusion coefficients for liquids lower than 10^4 to 10^5 times.

One of the earliest equations for determining the diffusion coefficient in dilute solutions was the *Stokes-Einstein equation*, based on the model of motion of a spherical particle of diffusing substance A in a viscous liquid continuum B

$$D_{AB} = \frac{kT}{b\pi r_o \eta_B},$$

where r_o is the particle (molecule) radius and η_B, the liquid viscosity. The constant b depends on the size of diffusing molecules: b = 6 for molecules larger than those of the base substance; b = 4 for identical molecules; and b can be less than 4 for smaller molecules.

Assuming that the molecule diameter $2r_o = (\bar{v}/N_o)^{1/3}$, where $\bar{v}$ is the molar volume of a diffusing substance and N_o is the Avogadro number, then

$$D_{AB} = \frac{2kT}{b\pi\eta_B}\left(\frac{N_o}{\bar{v}_A}\right)^{1/3}.$$

Comparison of this formula with experimental data has shown that in most cases, the discrepancy is moderate and reaches 40% only in some cases.

In 1955, *Wilke and Chang* have suggested a more general formula based on extensive experimental investigations, but involving many empirical values as well

$$D_{AB} = 7.410^{-8}\left[(\varphi\bar{\mu}_B)^{1/2}\frac{T}{\eta_B \bar{v}_A^{0.6}}\right],$$

where D_{AB} is the interdiffusion coefficient in an infinitely-dilute solution, cm^2/s; φ, the parameter of association of solvent B; $\bar{\mu}_B$, the molecular mass of substance B; $\bar{\mu}_A$, the molar volume of solute A at a boiling point under normal conditions, cm^3/mol; η_B, the substance viscosity, Ns/m^2; and T, the temperature, K.

Introduction of the association parameter into the formula is brought about by the fact that associated molecules behave like large-size molecules and diffuse at a lower rate; the degree of association varying with mixture composition and with molecule types. Therefore, Wilke and Chang presented the values for most widespread solvents: for water φ = 2.6; methanol, 1.9; ethanol, 1.5; benzene, ester, heptane and nonassociated solvents, 1.

A semiempirical formula suggested by *Scheibel*,

$$D_{AB} = 8.210^{-8} T\left[1+\left(\frac{3\bar{v}_B}{\bar{v}_A}\right)^{2/3}\right]\eta_B^{-1}\bar{v}_A^{-1/3},$$

is worthy of attention. In some cases, it appears to be more exact than the preceding one; but for $v_A/v_B \leq (1 - 2)$, the deviation from experiment becomes important and the following relations are recommended:

$D_A = 25.210^{-8} T\eta_B^{-1}\bar{v}_B^{-1/3}$ (water is the solvent);

$D_A = 18.610^{-8} T\eta_B^{-1}\bar{v}_B^{-1/3}$ (benzene is the solvent);

$D_A = 17.510^{-8} T\eta_B^{-1}\bar{v}_B^{-1/3}$ (other solvent).

Reddy and Doraiswamy have suggested the equation

$$\frac{D_{AB}\eta_B}{T} = K_{RS}\frac{\bar{\mu}_B^{1/2}}{(\bar{v}_A\bar{v}_B)^{1/3}},$$

where K_{RS} varies depending on the ratio of molar volumes: $K_{RS} = 8.5\ 10^{-8}$ for $\bar{v}_A/\bar{v}_B > 1.5$ and $K_{RS} = 10^{-7}$ for $\bar{v}_A/\bar{v}_B < 1.5$. Comparison of this equation with 96 experimental points has shown good agreement, the spread of points being about +15%.

All formulas suggested above for calculating the diffusion coefficient hold true for low-viscosity liquids. For a high-viscosity solvent, they are in great error and therefore inapplicable.

The temperature effect on the diffusion coefficient has been poorly studied so far. Within a narrow temperature range—from 10 to 20°C—the temperature dependence of the diffusion coefficient can be assumed to be linear

$$D = D_{298}[1+\alpha\langle T-298\rangle],$$

where D_{298} is the diffusion coefficient at T = 298K, $\alpha = 20\eta_{298}^{1/2}\rho^{-1/3}$; η_{298}, the solvent viscosity at T = 298K, Ns/m; ρ, the solvent density, g/cm³.

Theoretically, this must be an exponential dependence of the type D = AT exp (-E/RT).

The experimental data of Wilke and Chang give available evidence that the activation energy varies from 12.6 to 28.1 kJ/mol.

The dependence of the diffusion coefficient on concentration of diffusing substance, strictly speaking, is a consequence of the fact that diffusion flow depends on the difference (gradient) of the thermodynamic potential of the system rather than concentration, i.e., the formula must allow for activity of the diffusing substance. Hence, at V_B = const,

$$D = D_0 \frac{d\ln a}{d\ln c} D_0\left(1+\frac{d\ln v}{d\ln c}\right),$$

where D_0 and D are the diffusion coefficients, respectively, in an infinitely-dilute solution and in a solution with finite concentration c; a and c, the activity and the concentration of diffusing substance; and n, the activity coefficient of this substance. The semiempirical formulas presented above are more exact than the theoretical ones because the latter were derived making assumptions. Nevertheless, to avoid an appreciable error it is advisable to make calculations by several formulas concurrently and to compare the results.

In electrolyte solutions, salts dissociate and diffuse as ions and molecules depending on the degree of dissociation. The theory of salt diffusion is elaborated mainly for dilute solutions in which the degree of dissociation is close to one. Thus, the diffusion coefficient for a simple salt that is infinitely diluted can be found using the *Nernst-Heckell equation*

$$D_{AB} = \frac{RT}{Fa^2}\frac{1/n_+ + 1/n_-}{1/\lambda_+^0 + 1/\lambda_-^0}$$

where D_{AB} is the diffusion coefficient, defined as the proportionality factor between the molecular flow of dissolved salt and the gradient of its molecular concentration, cm²/s; T, the temperature, K; Fa, the Faraday number, n+ and n-, the cation and anion valences; λ_+^o and λ_-^o , the limit (under an infinite dilution) ionic conductions of cation and anion, cm²/Ω mol.

In electrolyte solutions, the diffusion coefficient substantially depends on the concentration of diffusing substance. If its concentration is no more than 2N, the formula

$$D = D_0 \frac{V}{x_1 V_1} \frac{\eta_1}{\eta} \left(1 + m \frac{d \ln v}{dm} \right)$$

is suggested by Gordon, where D_0 and D are the diffusion coefficients for an infinitely dilute solution and a molar solution, respectively; V, the solution volume; V_1, the partial volume of the solvent; x_1, the number of solvent moles in volume V; v, the molarity of the solution; η and η_1, the viscosity of solution and solvent; v, the molar activity coefficient.

References

- Fr. 12; see pp.291–2 of Kirk, G. S.; J. E. Raven; Malcolm Schofield (1983). The Presocratic Philosophers (2 ed.). Cambridge: Cambridge University Press. ISBN 978-0-521-27455-5
- Whitaker, Robert D. (1975-10-01). "An historical note on the conservation of mass". Journal of Chemical Education. 52 (10): 658. Bibcode:1975JChEd..52..658W. doi:10.1021/ed052p658. ISSN 0021-9584
- Robert D. Whitaker, "An Historical Note on the Conservation of Mass", Journal of Chemical Education, 52, 10, 658-659, Oct 75
- Weber, Walter J., Jr. (1972). Physicochemical Processes for Water Quality Control. Wiley-Interscience. ISBN 0-471-92435-0

Permissions

All chapters in this book are published with permission under the Creative Commons Attribution Share Alike License or equivalent. Every chapter published in this book has been scrutinized by our experts. Their significance has been extensively debated. The topics covered herein carry significant information for a comprehensive understanding. They may even be implemented as practical applications or may be referred to as a beginning point for further studies.

We would like to thank the editorial team for lending their expertise to make the book truly unique. They have played a crucial role in the development of this book. Without their invaluable contributions this book wouldn't have been possible. They have made vital efforts to compile up to date information on the varied aspects of this subject to make this book a valuable addition to the collection of many professionals and students.

This book was conceptualized with the vision of imparting up-to-date and integrated information in this field. To ensure the same, a matchless editorial board was set up. Every individual on the board went through rigorous rounds of assessment to prove their worth. After which they invested a large part of their time researching and compiling the most relevant data for our readers.

The editorial board has been involved in producing this book since its inception. They have spent rigorous hours researching and exploring the diverse topics which have resulted in the successful publishing of this book. They have passed on their knowledge of decades through this book. To expedite this challenging task, the publisher supported the team at every step. A small team of assistant editors was also appointed to further simplify the editing procedure and attain best results for the readers.

Apart from the editorial board, the designing team has also invested a significant amount of their time in understanding the subject and creating the most relevant covers. They scrutinized every image to scout for the most suitable representation of the subject and create an appropriate cover for the book.

The publishing team has been an ardent support to the editorial, designing and production team. Their endless efforts to recruit the best for this project, has resulted in the accomplishment of this book. They are a veteran in the field of academics and their pool of knowledge is as vast as their experience in printing. Their expertise and guidance has proved useful at every step. Their uncompromising quality standards have made this book an exceptional effort. Their encouragement from time to time has been an inspiration for everyone.

The publisher and the editorial board hope that this book will prove to be a valuable piece of knowledge for students, practitioners and scholars across the globe.

Index